Sustainability

The concept of sustainability is traditionally viewed in exclusively environmental terms. *Sustainability: Life Chances and Livelihoods* links peoples' livelihoods and life chances to the concept of sustainability by examining the way in which social and economic processes complement and compound environmental change. Looking at the main ingredients of sustainable development – health, economic policy, land use, ethics and education, in both the north and south, this book demonstrates the way in which the life chances of individuals both affect and are affected by, their environments.

Sustainability: Life Chances and Livelihoods shows that the scope of sustainability thinking needs to be widened to embrace public policies and experiences in both developed and developing countries. By providing a comparative focus, both spatially and temporally, the contributors demonstrate how the environmental concerns of the northern developed world are culturally translated into the south, often into immediate survival questions.

Michael Redclift was formerly Professor of International Environmental Policy at Keele University. He is now Professor of Human Geography and the Environment, King's College, London University.

Sustainability

Life chances and livelihoods

Edited by Michael Redclift

London and New York

First published 2000
by Routledge
11 New Fetter Lane, London EC4P 4EE

Simultaneously published in the USA and Canada
by Routledge
29 West 35th Street, New York, NY 10001

Routledge is an imprint of the Taylor & Francis Group

Typeset in Galliard by RefineCatch Limited, Bungay, Suffolk
Printed and bound in Great Britain by
MPG Books Ltd, Bodmin

British Library Cataloguing in Publication Data
A catalogue record for this book is available from the British Library

Library of Congress Cataloging in Publication Data
Sustainability : life chances and livelihoods / edited by Michael
    Redclift.
        p.   cm.
    Includes bibliographical references and index.
    1. Sustainable development.   2. Economic development –
Environmental aspects.   3. Economic development – Social aspects.
    I. Redclift, M. R.
    HC79.E5S86654   1999
    338.9′26 – dc21   99–29732

ISBN 0–415–19617–5 (hbk)
ISBN 0–415–19618–3 (pbk)

Contents

List of figures   vii
List of tables   viii
List of contributors   ix

1   Introduction   1
    MICHAEL REDCLIFT

PART 1
The environment and public policy   15

2   Sustainability, knowledge, ethics and the law   17
    JOHN PROOPS AND DAVID WILKINSON

3   Environmental policy-making: what have economic analysis and
    the idea of sustainability got to offer?   35
    PAUL EKINS

4   Land use policy and sustainability   58
    P. T. KIVELL

5   Environmental health and sustainability   75
    K. T. MASON

PART 2
Historical perspectives on sustainable livelihoods   93

6   Population, food and agriculture in mid-nineteenth century England   95
    A. D. M. PHILLIPS

7   Envisaging the frontier: land settlement and life chances in
    Upper Canada   106
    MICHAEL REDCLIFT

PART 3
Geographical perspectives – the view from the South 121

8 Exploring dimensions of sustainability in Nigeria: a question of scale 123
ELSBETH ROBSON

9 Sustainability: life chances and education in Southern Africa 144
NICOLA ANSELL

10 Linking the past with the future: maintaining livelihood
strategies for indigenous forest dwellers in Guyana 158
CAROLINE SULLIVAN

Index 189

Figures

6.1   Agricultural land use in Staffordshire, *c*. 1840–70   99
6.2   Agricultural rent per acre in Staffordshire, 1842–69   103
8.1   Hausaland   126
8.2   Location of Zarewa Village, Kano State, Northern Nigeria   127
8.3   Climate, seasons and agricultural activities in Hausaland   129
8.4   Primary school admissions in Zarewa   135
8.5   The condition of Nigerian education   136
9.1   Economic integration of the labour reserve   146
10.1 The location of the study villages in North-West Guyana   165
10.2 A comparison of the average household values of productive
      outputs in three Amerindian villages in North-West Guyana   173
10.3 Productive wealth holdings in three Amerindian villages in
      Guyana, 1996   180

Tables

3.1   Matrix for the construction of the Sustainability Gap   51
4.1   Projections of urban growth, by county (England), 1991–2016   64
6.1   Price indices of butter, beef, mutton and wheat in Staffordshire
      markets, 1840–70   97
6.2   Agricultural land use in Staffordshire, 1840–70   100
6.3   Cropping in Staffordshire, 1840–70   100
6.4   Density of dairy cattle in eight Staffordshire parishes, 1840–70   101
8.1   Nigerian economic statistics, 1985–98   138
9.1   Percentage of pass rates, MHS and RSS   149
10.1 Characteristics of study villages   164
10.2 A comparison of total crop output in three Amerindian villages,
      Guyana, 1996 (1)   166
10.3 A comparison of total crop output in three Amerindian villages,
      Guyana, 1996 (2)   167
10.4 Crop prices in Assakata, Sebai and Karaburi, 1996   168
10.5 Effective labour supply   169
10.6 A comparison of total and average labour input values, by
      sector, in three Amerindian villages in Guyana, 1996   170
10.7 Meat and fish price variation across villages, 1996   171
10.8 Productive sectoral output values for three Amerindian villages
      in Guyana, 1996   172
10.9 A comparison of bushmeat catches and household consumption
      in three Amerindian villages in North-West Guyana, 1996   173
10.10 A comparison of fishing catches and household consumption in
      three Amerindian villages in North-West Guyana, 1996   174
10.11 A comparison of the net sectoral value of forest inputs   176
10.12 Capital availability, 1996   179
10.13 Palm-heart harvesting in Guyana, 1990–95   181
10.14 The relative importance of development issues, as indicated by
      mean scores assigned by men and women   184

Contributors

Nicola Ansell is Temporary Lecturer in the Department of Environmental Social Sciences (ESS), Keele University.

Paul Ekins is Reader in Environmental Policy, ESS Keele University.

Philip Kivell is Reader in Human Geography, ESS Keele University.

Keith Mason is Senior Experimental Officer, ESS Keele University.

Anthony Phillips is Reader in Historical Geography, ESS Keele University.

John Proops is Professor of Ecological Economics, ESS Keele University.

Michael Redclift was Professor of International Environmental Policy, ESS Keele University. He is now in the Geography Department, King's College, London.

Elsbeth Robson is Lecturer in Development Studies, ESS Keele University.

Caroline Sullivan was formerly a graduate student at Keele University and currently works at the Institute of Hydrology.

David Wilkinson is Lecturer in Environmental Law, ESS Keele University.

1 Introduction

Michael Redclift

Introduction

The concept of sustainability has often been used in a rather cavalier fashion: for example, to demonstrate philosophical points about *all* societies, without grounding the observations in historical time, or to contrast *other* societies with our own, thereby urging a political agenda through implied comparison. The concept of sustainability has rarely been advanced by such sweeping, and ahistorical, comparisons. In particular, there has been a woeful neglect of what sustainability might mean for other people, in other cultures, or in other times.

In this book we want to illuminate practices, as well as analytical understanding; and to locate sustainability (and its absence) in everyday life. We begin by examining the use to which sustainability is put in the public policy discourses that surround us in the developed countries. This becomes our point of departure, but these experiences need to be linked to two other dimensions of sustainability. The first is the temporal dimension – the historical context in which communities maintain, or defend, their cultural and economic integrity. The second concerns livelihoods in developing countries today – where sustainability is often not a prescription for 'alternative' Green values, but a defence of existing values, tied to specific forms of reproduction and behaviour, under threat from external economic forces, or the collapse of a delicate natural resource balance.

The interest in sustainability in the developed world is linked to two currents which have come to characterise the interface between human aspirations and the mastery of nature. The first process involves finding economic values or indicators for economic and social phenomena – the economisation of society. Human activities are translated into economic terms, and removed from both their environmental and cultural contexts. The second process involves a revaluation of nature – nature becomes 'socialised' – and transformed into something which can be managed and controlled, which can be evaluated through quantitative indices. Both currents are represented in the term 'sustainable development', which embodies the contradiction between human aspirations for domination over nature, and our ultimate dependence on natural systems and ecological constraints. Modernism has been characterised as a discourse predicated upon the dualism of nature and culture, and as such a denial of the essentially 'social' character of nature (Braun and Castree 1998). As we shall see later, both processes, the economisation of society and the socialisation of nature, raise ethical, distributive and 'rights' issues. The discussion of both processes leads us towards a redefinition of citizenship itself.

Sustainable development

Each scientific 'problem' that is resolved by human intervention, using fossil fuels or new materials and technologies, is often viewed as a triumph of management, and a contribution to economic good, when its 'solution' might also represent a future threat to sustainability. Having jettisoned the fear that resources themselves were limited, in the 1970s, we are today faced by the prospect that the means we have used to overcome resource scarcity, resource substitution and increased levels of industrial metabolism, themselves contribute to the next generation of problems that are associated with the global environment. This realisation provides an enormous challenge to conventional social science thinking, a challenge encapsulated in the term 'sustainable development'.

Sustainable development was defined by the Brundtland Commission in the following way: 'development that meets the needs of the present without compromising the ability of future generations to meet their own needs' (Brundtland 1987). This definition has been brought into service in the absence of agreement about a process which almost everybody thinks is desirable. However, the simplicity of this approach obscures underlying complexities and contradictions. Before exploring, later in this volume, whether we can establish indicators of sustainability, it is worth pausing to examine the apparent consensus that reigns over sustainable development.

First, following the Brundtland definition, it is clear that 'needs' themselves change, so it is unlikely (as the definition implies) that those of future generations will be the same as those of the present generation. The question then is, where does 'development' come into the picture? Obviously development itself contributes to 'needs', helping to define them differently for each generation, and for different cultures. This focus on what distinguishes the needs of different societies is represented in parts 2 and 3 of this book.

There is another, second, question, not covered by this definition of sustainable development, which concerns the way in which needs are defined in different cultures. Most of the 'consensus' surrounding sustainable development has involved a syllogism: sustainable development is necessary for all of us, but it may be defined differently in terms of each and every culture. This is superficially convenient, until we begin to ask how these different definitions match up. If in one society it is agreed that fresh air and open spaces are necessary before development can be sustainable, it will be increasingly difficult to marry this definition of 'needs' with those of other societies seeking more material wealth, even at the cost of increased pollution. And how do we establish which course of action is *more* sustainable? Recourse to the view that societies must decide for themselves is not very helpful. (Who decides? And in whose interests? On what basis are the decisions made?) At the same time there are problems in ignoring culturally specific definitions in the interest of a more inclusive ontology.

There is also considerable confusion surrounding *what* is to be sustained. One of the reasons why there are so many contradictory approaches to sustainable development (although not the only reason) is that different people identify the objects of sustainability differently.

For those whose primary interest is in ecological systems and the conservation of natural resources, it is the natural resource base which needs to be sustained. The key question that is usually posed is the following: how can development activities be designed which help to maintain ecological processes, such as soil fertility, the assimilation of wastes, and water and nutrient recycling? Another, related, issue is the conservation of genetic materials, both in themselves and (perhaps more importantly) as part of complex, and vulnerable systems of bio-diversity. The natural resource base needs to be conserved because of its intrinsic value.

There are other approaches, however. Some environmental economists argue that the natural stock of resources, or 'critical natural capital', needs to be given priority over the flows of income which depend upon it. They make the point that human-made capital cannot be an effective substitute for natural capital, the claim of so-called 'weak' sustainability (Pearce 1993: 15–17). If our objective is the sustainable yield of renewable resources, then sustainable development implies the management of these resources in the interest of the natural capital stock. This raises a number of issues which are both political and distributive: who owns and controls genetic materials, and manages the environment? At what point does the conservation of natural capital unnecessarily inhibit the sustainable flows of resources? Second, according to what principles are the social institutions governing the use of resources, organised? What systems of tenure dictate the ownership and management of the natural resource base? What institutions do we bequeath, together with the environment, to future generations? Far from taking us away from issues of distributive politics, and political economy, a concern with sustainable development inevitably raises such issues more forcefully than ever.

The question 'what is to be sustained?' can also be answered in another way. Some writers argue that it is present (or future) levels of production (or consumption) that need to be sustained. The argument is that the growth of global population will lead to increased demands on the environment, and our definition of sustainable development should incorporate this fact. At the same time, the consumption practices of individuals will change too. Given the choice, most people in India or China might want a television or an automobile of their own, like most households in the industrialised North. What prevents them from acquiring one is their poverty, their inability to consume, to exert their influence on the market, as well as the relatively 'underdeveloped' infrastructure of poor countries.

Is there anything inherently unsustainable in broadening the market for computers, TV sets or cars? If the answer is 'yes', then those of us who possess these goods need to be clear about why we consume goods unavailable to others. The response is usually that it is difficult, or even impossible, to function in our society without information or private motorised mobility. But, this is to evade the question of what I would call 'underlying social commitments': the taken for granted aspects of everyday social life (Redclift 1996). We may define our needs in ways which effectively exclude others meeting theirs, and in the process increase the long-term risks for the sustainability of their livelihoods. Most importantly, however, the implications of the processes through which we enlarge our choices, and reduce those of others, is largely invisible to us, it may take place at several removes from us, in other countries, or in the future.

If we concentrate our attention on our own society, we can begin by identifying aspects of our management of the environment that are unsustainable. It is a short step, as we shall see, to the development of sustainability indicators. The growth of interest in sustainability indicators has followed that of sustainable development. Again, the importance of this issue is matched by the difficulty in addressing it convincingly, as Paul Ekins argues in Chapter 3. There are numerous indicators of *un*sustainability, but it has proved much more difficult to find those for sustainability.

The reasons for these difficulties are not hard to find. Economics developed, historically, around the idea of scarcity; it was the 'dismal science'. The role of technology was principally that of raising output from scarce resources. Among other benefits of economic growth was the political legitimacy it conferred, within a dynamic economy, on those who could successfully overcome the obstacles to more spending. Wealth was regarded as a good thing, in itself. This proposition, which underlines the difficulty in reconciling 'development' with 'sustainability', strikes at the legitimisation of only one form of 'value' within capitalist, industrial societies. Habermas expressed his criticism of this view forcefully, in the following way:

> can civilisation afford to surrender itself entirely to the . . . driving force of just one of its subsystems – namely, the pull of a dynamic . . . recursively closed, economic system which can only function and remain stable by taking all relevant information, translating it into, and processing it in, the language of economic value.

> (Habermas 1991)

This issue, of how we value the environment, and the implications of our valuation for planning sustainability, is addressed at several points in this book, but particularly in the chapters by Paul Ekins and Caroline Sullivan.

The environment and public policy

It is clear that once we begin to prise behind surface appearances 'sustainability' means different things to different people. To some the 'environment' is linked indelibly with their livelihoods, with the way they make a living. To others the environment suggests a place, or an activity, which is associated with the time they spend on recreation – it is essentially non-vocational. These different kinds of emphasis are sometimes, wrongly in our judgement, associated with poverty and affluence, respectively. The industrialised North, it is said, is characterised by a concern with *lifestyles*, implying the importance of taste, fashion and personal consumption, all expressed through the ability to exercise choice in the market-place. This is sometimes compared, often in sweeping terms, with the *livelihoods* that are depicted in the poorer countries of the South, where people eke out a bare living, against a background of resource degradation and poverty.

This way of dividing up the globe is both inaccurate and confusing. The pursuit of goods, of personal identity, even of fashion, is not confined to the rich, or to the rich worlds. It is also exhibited in squatter settlements and popular culture, in urban *barrios* and rural villages throughout the developing world (as well as in ethnic minority communities in the North). In the same way, the effect of the environment in underpinning peoples' life chances, in helping to determine their health or financial security, is not confined to the poor, or to the industrialised world, as Keith Mason argues in this volume, where you live, even in developed countries, is quite a good indicator of the quality of your health, and your life chances in general. In reality we are subject to pressures – in livelihoods and lifestyles – wherever we live, and whatever our relationship to the environment.

The direction that environmental policy is taking, particularly in the North, is linked to the 'balance' that societies achieve in seeking both economic growth and environmental protection. In Chapter 4 Philip Kivell examines the demands of space – especially urban space – on limited physical resources, and apparently irreconcilable planning objectives. To fully appreciate the different ways in which environmental risks and benefits are manifested, we need to identify different spheres of activity, in all of which sustainability is a key organising principle. As argued in several of the chapters in this volume (perhaps most notably by Mason and Kivell) sustainability is most usefully employed as an integrative concept, serving to integrate important policy domains, such as health and land use planning, within the wider compass of public policy.

One way of conceptualising these distinctions, and to integrate sustainability as a concept, is to identify different spheres of environmental activity. These differ-ent spheres of activity in which sustainability can be analysed are as follows.

Spheres of environmental activity

1   The sphere of production

- the individual's immediate work environment (industrial risks, plant pollution)
- workface conflicts: health, work-related risks
- indirect consequences of production activities: waste, toxins, pollution

2   The sphere of consumption

- the individual's consumption practices and choices: health risks, food risks

- indirect consequences of consumption: food miles, ghost acres, ecological footprints
- socially generated consumption: energy, waste

3   The sphere of social capital/infrastructure

- the built environment, urban space, public utilities, transport infrastructures
- spatial structures and access to services with distributional consequences (collective consumption)

4   The sphere of 'nature'

- amenity and the countryside
- positional goods and landscape
- access to wilderness
- animal rights and welfare

5   The sphere of physical sustainability

- air pollution, climate change
- ozone depletion
- stability of coastlines, watershed forests, human management of 'natural disasters'

Environmental policy assumes a series of background activities. It follows that we need to consider the wider relationship between society and nature, within which policies and technologies are developed. As the diagram above shows, we can identify five different spheres of environmental activities, all of which carry associated benefits and costs, and risks associated with environmental uses.

The first sphere is that of production. The environmental effects of production are identifiable in a numbers of ways: through the immediate work environment, such as productive plant; through the effects of employment on health and welfare, and in meeting the risks associated with industrial activity. There are also important indirect effects of production activities, that are not so geographically confined, and may contribute to other spheres – such as transnational pollution, waste disposal or nuclear radiation.

Another sphere of environmental activity is that of consumption. This corresponds to the individual's consumption of goods and services, and the risks associated with these consumption activities. As in the case of production, there are also indirect dimensions of consumption to be considered: the effect of consumption in one part of the planet on other areas, often distant in geographical terms. This concept of *displaced* consumption is captured in terms like food miles (the transport miles associated with delivering food to the consumer) and ecological footprints (the impact on the environment of geographically remote areas of colonial or post-colonial trade and exchange) as materials and natural resources are sourced from far away. Finally, the sphere of consumption includes socially generated consumption – the generation of energy to meet consumer demand, and the disposal of waste and packaging to meet these demands.

The sphere of social capital and infrastructure provides the physical context against which the individual works and plays: the built environment, urban space, utilities such as energy, water and waste disposal, and the transport systems that link these activities. Under this heading we should also consider public services which are provided on a collective basis – such as public parks, recreational services and open areas.

The fourth sphere of environmental activity is represented by 'nature', in the sense of life-forms and biological systems external to human being, the country-side, forests and landscape. This sphere does not only include physical manifest-ations of the natural world, which vary in value according to the level of public access ('positional goods' to Hirsch 1976) but also the nature of sentient beings, animals and birds. Within the sphere of nature, then, we should also consider the rights and obligations of animal life, and our rights and obligations *to* animals, and their welfare.

The final sphere of environmental activity might be termed that of *physical* sustainability, in the sense of background processes which govern the quality of the total environment, like climate change, the depletion of the ozone layer, and destruction of forests and watershed basins through anthropogenic causes. Against the backdrop of shifts in physical sustainability other human activities often serve to exacerbate environmental degradation: the destruction of wilder-ness areas, or national parks, for example, often takes place against a background of acid rain, and may contribute to global warming.

Clearly, then, these spheres of activity, in which human actions strengthen, or undermine, sustainability, and the quality of both human and natural capital, are closely inter-related. They are separate spheres only in the analytical sense that we can separate their causes and consequences, and that they contribute to under-mining sustainability, by increasing the vulnerability of both human populations and ecological systems. Human societies transform the environment through processes (*taking, adding and replacing* according to Dunlap 1992) which cor-respond with the different uses to which the environment is put. These activities, taken together, correspond to different facets of human behaviour, and to the way that societies are organised.

One clearly demonstrable area in which environmental policy is being developed at the European level is that of technological policy, and changes in production processes which seek to internalise environmental 'externalities'. But this process – often described as ecological modernisation (Gouldson and Murphy 1998) – is only part of a much larger picture. As we shall see, to fully understand the public policy discourses which serve to frame environmental problems, we need to be aware of the effect of environmental uses – material flows and energy use – on the way that societies are organised. One approach, following the diagram above, would be to link sustainability with the different domains in which recourse is made to the environment.

Existing ways of accessing the environment, and transforming it, reflect systems of power, and the exercise of power through *choices*, many of them market choices, which themselves are dictated by the distribution of income, and the existence of different social classes. Environmental policy and management is not divorced from a social context; it is the *product* of a social context. Environmental concerns are not only the outcome of the individual's immediate experience of production and consumption, they are also socially constructed, and reflect the institutions which govern our behaviour and values. The way in which environmental problems, and policies, are constructed in societies influences both the *scope* for achieving sustainability, and the *way* in which sustainability can be achieved.

The chapters in Part 1 of this book draw together a series of related concerns:

- The relationship between economic efficiency and environmental benefits – what is most effective, and what is ethical, as Proops and Wilkinson put it in Chapter 2.
- The interest in both current inequalities (*intragenerational equity*) and differences between this and future generations (*intergenerational equity)*.
- The integration of sectoral policies – for health or land use – with the wider dimensions of sustainability planning which (as Kivell notes in Chapter 4) is understood differently by different people.

In their chapter Proops and Wilkinson begin with the 'building blocks' of sustainability – environmental knowledge and values. They point out that, in terms of public policy, the environment is particularly challenging. There is, on the one hand, considerable uncertainty associated with environmental risks and problems. This may be partly accounted for by the inadequate understanding of environmental issues by the 'general public', but since there are clearly distinguishable 'publics', it also arises because of different interests in the environment. Often 'public' and 'expert' understandings of the environment are at variance with each other (Beck 1992). The growth of what has been termed 'expert knowledge' implies, in its turn, a growth in 'ignorance' from non-experts, whose *interests* in the environment are just as important as those of specialist advisors and scientists. Proops and Wilkinson discuss the way in which environmental knowledge and environmental values are related. They end by considering the way in which the law might influence environmental ethics and, in turn, be transformed by more attention to the issues surrounding sustainability.

The following chapter, by Paul Ekins, explores the need for sustainability goals and sustainability standards for public policy. Ekins distinguishes three key elements of sustainability, which each challenge public policy. These are: the problems of future generation; those of distinctive interests in the environment (victims and beneficiaries of the environment); and the rights of non-human species. He considers whether economics can square up to the challenge of developing effective policy to address these issues. By looking at *un*sustainable environments, at the loss of important environmental functions, he seeks to develop an improved basis for sustainable policy. Like other contributors to this volume, Ekins believes that sustainability cannot be addressed exclusively by 'environmental' policy – what is required is economic and social progress that is also sustainable progress. Although, in his view, economics has an important role to play in charting environmental progress, and for assessing economic costs and benefits, the eventual goal of sustainability cannot be achieved by economic deliberation alone.

The next chapter, by Philip Kivell, examines the sustainability debate from the perspective of land use policy. Land is clearly of critical importance to any debate about sustainability, since it is an essential ingredient of all human activity, and an element in most of the spheres of production/consumption outlined above. A concern with land use, much of which originated in the United Kingdom, is also central to the process of planning; land use controls help promote and regulate both economic development and environmental protection. Increasingly, moreover, this model which places land use at the centre of the frame, is being used to buttress environmental management interventions at the international, and global, levels. In his chapter Philip Kivell reviews a number of aspects of urban form and process, all of which require consideration of 'sustainability'; in one guise or another. They also lead us to consider the importance of mixed urban functions in planning for more sustainable cities.

The last chapter in Part 1, by Keith Mason, looks at the role of environmental health in planning for sustainable development. Mason argues that human health – both physical and psychological – is closely related to the 'variable quality of the local environment', and goes on to distinguish three important sets of relationships governing sustainability and health: health and the environment, the environment and economy, and the economy and health. The chapter goes on to examine the role of contextual environmental differences in determining life chances. It also assesses the links between health and the environment through a review of interpretative tools, and the use of indicators, in compiling health profiles. Drawing on case material from North Staffordshire, the chapter demonstrates not only the way in which environmental effects are geographically distributed (through using Geographical Information Systems) but – in broader terms – the links between traditional and 'modern' health hazards and risks.

Sustainable livelihoods in historical perspective

Part 2 takes the discussion of life chances and livelihoods in a different direction, by comparing historical evidence. It examines the evidence for the maintenance and reproduction of livelihoods in rural areas of England, and Canada, and examines the environmental impacts of this process in each of these anglophone geographical areas. Many of the issues raised in the first part of this book surface again, often in ways that force one to re-examine assumptions. For example, what are the implications for *our* generation of the way our ancestors transformed their environments, and the values that these transformations gave rise to? To what extent does the experience of societies whose livelihoods were based on local resources, and locally available labour, provide lessons on sustainability for 'us' today? In thinking about sustainability in the context of public policy in the industrialised world we often forget that for most people in the past, sustainability was circumscribed by relatively narrow geographical boundaries. Sustainability, in the context of rural communities in the past, was about the social institutions that managed local resources. These conditions, as we shall see in Part 3 of the book, also characterise most of the world's population in the developing world today.

The chapters in Part 2 examine the links between land, food and population, emphasising the continuities – and discontinuities – experienced by communities, and their struggles for livelihood. Sustainability, in the context of such communities, is concerned with securing, and protecting, a 'way of life' over time, and often in the face of external threat. Sustainability is not merely about how nature is constructed socially, it is also about the reproduction of social organisations and institutions.

The two chapters on sustainable livelihoods from an historical perspective, also examine the balance that is struck between demographic factors and the way natural resources, especially land, are used and managed. The need to meet livelihood possibilities – particularly for food and employment – is considered in relation to the structure of rural communities. Changes in the structure of communities over time prompts reflections on the necessary conditions for securing local livelihoods, and their social reproduction. This temporal dimension to the analysis of sustainability, which was touched on in the earlier chapter by Ekins, can inform the wider discussion, through a close examination of the *institutions which we bequeath subsequent generations.* Being able to arrive at sustainable levels of exploitation in *environmental* terms, by conserving resources and enhancing 'sink' capacities, tells us nothing about the social and economic mechanisms that will be required to achieve these objectives. By grounding the discussion of sustainability in the lived experience of people in rural communities, we can appreciate the primacy of these social mechanisms, without which the 'environment' is an abstraction.

The chapter by Anthony Phillips (Chapter 6) looks at the relationship between population and agricultural change in England in the nineteenth century. It concentrates on the way in which changes in land tenure, and innovations in agricultural technology, provided the basis for new 'levels' of sustainable production. At the same time, the systems of agricultural production in England in this period established the 'norms' which we have inherited and, at several historical removes, have served to contribute to many of the problems surrounding sustainability in England today.

The prospect of gaining land, and the inducements provided by the market, to cultivate it non-sustainably, is also a key element in the chapter by Michael Redclift, with which this section of the book ends. In societies that have been settled for long periods of time, such as those in the British Isles, the tension between livelihood and economic development reflects attempts to achieve security for the rural household, in the face of hostility from other social classes, and oppressive social institutions. The settlement of North America, in particular, suggests a different set of tensions. Here the immigrant population – much of which was drawn from the British Isles – were propelled into migration by the prospect of freedom from economic and social constraints. The frontier communities of 'Upper Canada' in the 1840s, which provide the focus for this chapter, sought to establish their own institutions to manage land and other natural resources, in the absence of well-established social conventions. Under conditions of natural 'plenty', where labour was scarce and land plentiful, the migrant populations worked the land in ways that would not be considered 'sustainable' by many people today. These frontier communities saw the life chances of many of their members improve substantially, once they had survived the rigours of immigration. However, their view of nature, and the environment, was itself formed by their experiences of hardship, independence and confrontation with the wilderness they had settled. The legacy of their attempts to 'tame' this wilderness, to wrest livelihoods out of difficult natural conditions, and to establish a civil society, is reflected today in the enduring power of frontier myths and the refusal to acknowledge natural 'limits' to human ambitions. Many of the settler societies of North, Central and South America, as well as parts of Australasia and Southern Africa, have experienced similar experiences, establishing livelihoods at the expense of longer-term sustainability.

International perspectives on livelihoods and sustainability: developing country experiences

The chapters in Part 3 address the links between structures (particularly class and gender) and sustainability in developing countries, drawing on examples from Africa and the Caribbean. The focus, as in the previous section, is on sustainable *livelihoods*, and the ways in which resources are accessed by communities, processes which, in turn, play a large part in determining the household and communities' sustainability. In the context of developing countries a key concept in analysing life chances and livelihoods is that of *entitlements*, the entitlements enjoyed by individuals and groups with differential access to resources, and to structural power. Part 1 has discussed environmental preferences in the northern, industrialised world, where the language of market 'choice' is commonly used to establish individual preferences. In the context of developing countries, and especially in rural areas, the structural location of groups, defined by class, ethnicity and gender, is much more important in establishing their life chances, and the way in which they behave towards the environment.

The chapters in Part 3 examine the integrity of 'the household', and 'the community' as distinctive cultural settings for the discussion of sustainability and livelihoods. Case material is introduced to illustrate the way in which the social structure (marked by class, gender, ethnicity and age) intersects with that of the environment: facilitating, and denying, access to resources. Conflicts over resources, especially land and forests, in developing countries, thus challenges the institutional and legal framework of society. For most rural people these conflicts are largely around production (the domains of production and nature, in the diagram) but increasingly (as Nicola Ansell's chapter shows) these conflicts are also around the 'consumption' of services, such as secondary education.

It is sometimes suggested that a distinction needs to be made between environmental conflicts in the South, which usually seek to establish a new legal framework for settling environmental disputes, and those in the North, where the existing legal framework is invoked on behalf of environmental protestors. Such a distinction is crude, and somewhat inaccurate. In practice, many environmental conflicts in the developed world also seek to introduce a statutory framework which incorporates environmental objectives for the first time (as several chapters in Part 1 demonstrate), while environmental conflicts in the South invoke traditional institutional practices aimed at conserving resources. The three chapters in Part 3 underline, moreover, the important connections between struggles to achieve social and environmental objectives in developing countries, and the part played by livelihood creation in determining environmental goals.

The pursuit of more sustainable livelihoods in the South is bound up with the trade-offs between securing immediate needs – food, shelter and land – and ensuring that future generations inherit a sustainable resource base. Clearly, in the short term 'livelihood struggles' are often not 'sustainable', especially where the resource base is poor, and population pressures on it are increasing. However, the struggles for livelihood in the South do serve to remind us that whatever social mechanisms are employed by poor, rural people – migration, household survival strategies, or recourse to communal institutions as a safety net – the pursuit of social justice is intimately connected to rights to resources and the environment. Access to resources, and to the potential for sustainable livelihoods, is a citizenship issue in developing countries; just as the implications of reducing consumption, and enhancing sustainability in the North, have increasingly become merged with other demands from civil society.

All three chapters in Part 3 prompt questions about whether sustainability, in the context of the societies described, can be married with modernity. Can those who benefit from modern secondary education contribute to the enhanced sustainability of the rural communities from which they came? In northern Nigeria it is external economic pressures which have often served to undermine the sustainability of communities. As Elsbeth Robson points out, many Nigerians live in sustainable households and communities, within a national context that is clearly not sustainable. By contrast, in her chapter Caroline Sullivan points to ways in which local-level sustainability could be enhanced in tropical forest areas and, if built upon might point the way forward. In Guyana the interface with the global economy, and global culture, has placed the livelihoods of indigenous forest peoples in jeopardy. Can new forms of sustainable livelihoods be generated from this contact with planners and development agencies, that provide the individuals with better life chances, but do not threaten the integrity of local cultures? These are some of the issues which, as they emerge in developing countries, must make us reflect on the ethnocentric and ahistorical biases that have been brought to bear on the discussion of sustainability. It is clear that, by focusing on livelihoods and life chances, we are touching a rich seam of explanation, without which the idea of sustainability is a more impoverished, and limited, concept.

References

Beck, U. (1992) *Risk Society*, Sage, London.

Braun, B. and Castree, N. (1998) *Remaking Reality: Nature at the Millennium*, Routledge, London.

Brundtland Commission (World Commission on Environment and Development) (1987) *Our Common Future*, Oxford University Press, Oxford.

Dunlap, R. (1992) 'From Environmental to Ecological Problems', in C. Calhoun and G. Ritzer (eds) *Social Problems*, McGraw Hill, New York.

Gouldson, A. and Murphy, J. (1998) *Regulatory Realities: the implementation and impact of industrial environmental regulation*, Earthscan, London.

Habermas, J. (1991) 'What does socialism mean today?' in R. Blackburn (ed.) *After the Fall*, Verso, London.

Hirsch, F. (1976) *The Social Limits to Growth*, Routledge, London.

Pearce, D. (1993) *Blueprint Three: Measuring Sustainable Development*, Earthscan, London.

Redclift, M. (1996) *Wasted: counting the cost of global consumption*, Earthscan, London.

Part 1

The environment and public policy

2 Sustainability, knowledge, ethics and the law

John Proops and David Wilkinson

Introduction

There seems to be an emerging consensus that achieving sustainability will require substantial changes to human lifestyles and behaviour. These changes will need to be brought about through conscious policy, and be embodied in legal principles, at both the national and international levels. In this chapter we explore the ways we can hope to formulate such policies towards sustainability, when two issues are taken into account.

First, the nature and extent of our understanding of the environmental and social issues involved is problematic; we do not have firm knowledge foundations on which easily to construct effective and consensual policies.

Second, the policies we formulate will need to be more than just 'effective'. They will also need to reflect ethical principles, both towards humans and the rest of nature.

In seeking to understand how effective policies for sustainability can be constructed, we seek to engage with the issue of sustainability from three interlinked perspectives: knowledge, ethics and the law. We shall argue that while these three areas are often seen as being separate in their implications for human–nature interactions, they are best understood as being mutually supportive of an understanding of how problems of sustainability arise, and how policy may address these problems.

Conceptual background

Before we move on to examine the issues of knowledge, ethics and law with regard to the achievement of sustainability, it will be useful to sketch the conceptual framework that underpins our discussion. This can be summarised by six linked assumptions about humans, their effects on the world, their abilities, motivations and self-imposed restraints. First we state these assumptions briefly, then indicate how they generate the discussions in the rest of the chapter.

1   We live in a world where the damaging effects of human actions on nature are becoming increasingly evident.

2   Human abilities are steadily expanding, as our technical understanding increases, and is embodied in our modes of production and consumption.
3   Human desire for control and consumption seems unbounded. This leads to an ever increasing desire for the extension of human abilities.
4   Humans are ethical creatures, willing and able to distinguish, or at least debate, which actions are 'right' and which are 'wrong'. This leads to conflicts between human desires and codes of morality.
5   Human societies find it useful, even necessary, to bind themselves within frameworks of law. These limit the negative impact of human desires upon other members of society. In part, these legal frameworks reflect ethical concerns and understandings.
6   Human understanding is limited, but the nature of these limits is itself little understood.

In the light of the above six assumptions we can ask a range of interlinked questions, with corresponding tentative answers, which may guide us in our later analysis.

QUESTION 1: Do humans necessarily cause environmental damage?
TENTATIVE ANSWER 1: No. There are pre-modern societies which seem able to exist in reasonable harmony with their natural environments for considerable periods (and so are sustainable by definition).
QUESTION 2: Is modern technology necessarily destructive of nature?
TENTATIVE ANSWER 2: Possibly/probably not. There are emerging technologies which are less environmentally damaging than those they replace (cf. the 'Factor Four' debate and the notion of 'ecological modernisation' (von Weizsäcker *et al.* 1997)).
QUESTION 3: In the light of tentative answers 1 and 2, is the cause of current environmental damage the strength and nature of human desires?
TENTATIVE ANSWER 3: Yes, probably. Humans seem to strive for personal satisfaction through consumption (though they seem to be largely unsuccessful in this quest (Easterlin 1996: ch. 10)).
QUESTION 4: Do not humans recognise that such selfish behaviour, aimed only at achieving personal satisfaction, may be damaging to others, and that this is morally wrong?
TENTATIVE ANSWER 4: Yes. Moral understandings, embodied in moral precepts such as the Ten Commandments of the Christian Old Testament, seem common in human societies.
QUESTION 5: So is moral understanding the basis of law?
TENTATIVE ANSWER 5: Sometimes. One could argue that laws protecting life and property are at least partially moral in basis, though they are also clearly constructive of social well-being. Some laws, however, have no moral basis, and are only constructive of social well-being (e.g. traffic regulations).
QUESTION 6: Are laws to protect nature based on ethical considerations, or are they simply to protect social well-being?

TENTATIVE ANSWER 6: When the laws are to protect those alive today from the consequences of their own (social) actions, one could argue that a moral basis is not necessary; self-interest is sufficient explanation (e.g. laws on air pollution in cities). However, when laws seek to protect future generations or other species, it seems that ethical considerations must be involved (e.g. laws on climate change, or to protect the great whales).

QUESTION 7: If we have ethical considerations, which can be embodied in law, why do we still have environmental damage?

TENTATIVE ANSWER 7: There are two possible reasons:

1 our desire for consumption over-rides our ethical principles; and/or
2 we do not understand the world, or ourselves, sufficiently well to know how to implement effective policies which reflect our ethical concerns.

Having established a conceptual framework for our discussion, illustrated (we hope) with the above questions and tentative answers, we now turn to the problem of knowledge.

Knowledge and sustainability

We identify four areas of knowledge and understanding which are necessary for the establishment of well-formulated policies for sustainability.

1 Understanding of the natural world, and how our production and consumption activities impinge upon it.
2 Understanding of human perceptions and motivations, so we can know why we indulge in behaviour which is destructive to nature.
3 Understanding of ethical systems, so we can determine if human motivations which are destructive to nature might be morally constrained.
4 Understanding of the effectiveness of various systems of incentives and restrictions on human actions, so that appropriate restraining measures can be embodied in law.

We shall argue that all four areas are problematic, and impose significant difficulties in seeking to establish a sustainable world. We consider these four areas in turn. In this section we address the first two issues, while the ethical and legal issues are considered in later sections. (For an extensive discussion of problems of knowledge and ignorance in environmental policy analysis, see Faber *et al.* (1996: ch. 11).)

Understanding of the natural world

A necessary condition for the assessment of the sustainability of particular life-styles and activities is an understanding of the way human action impinges upon nature, and thereby upon social systems. The issue of human understanding is problematic on several levels. First, it is notoriously the case that environmental systems are complex and ill-understood, and we examine some illustrative issues.

Understanding and global warming

The debate over global warming is an extreme example of how difficult it is to know just what constitutes an environmental threat, the time scale of this threat, and the possibilities for countering it through policy action. The 'best' scientific evidence (Houghton *et al.* 1996) suggests that current trends in the emission of greenhouse gases will lead to global temperature rises in the range 0.5–2.5°C by 2050. The basic physics behind this prediction is rather simple and has been known for at least 100 years, depending only on the relative absorptive capacity of light by (principally) carbon dioxide, in the visible and infrared frequency ranges. (For an overview of global warming, see Leggett (1990).)

However, the uncertainties about the environmental effects from such warming are enormous. For example, increasing temperature may increase equatorial cloud formation, increasing the Earth's albedo (reflectivity), and thus inducing a nearly completely counteracting cooling effect. Rising temperatures may cause the major ice caps to begin to melt, raising sea levels. Alternatively, there may be increased precipitation at the poles, leading to the ice caps actually to increase in extent, so causing sea level fall. Increasing global temperature may cause the rapid release of methane from tundra and continental shelf areas, where it is found in enormous quantities as methane hydrite. As methane is a powerful greenhouse gas, this might lead to extreme temperature increases, perhaps high enough to raise the temperature of the deep oceans sufficiently to release some of the enormous quantities of carbon dioxide that reside there in solution. This could generate a runaway greenhouse effect, producing on Earth a climate akin to that of Venus.

On a more local level, any reduction of the northern ice caps could cut off the 'Atlantic conveyor', the deep ocean currents that generate the Gulf Stream and the North Atlantic Drift. This warm current keeps north-west Europe mild in winter, and its cessation would give this area a winter climate like that of Labrador, which is on the same latitude.

At present a major issue for global governance is finding appropriate responses to the huge, and hugely uncertain, threat of greenhouse-induced global climate change. Clearly, the ethical and legal issues involved in establishing a globally equitable and effective set of policies constitute an enormous challenge.

Understanding and ozone layer depletion

Another example of the problems of environmental knowledge is that of ozone layer depletion. While the potential dangers of the greenhouse effect could have been (and were) established a century ago, the problem of CFCs and the ozone hole was probably unpredictable. Indeed, the case of CFCs and ozone layer depletion is an example of the introduction of a material that, at first sight, had great health and environmental advantages, as it replaced highly toxic ammonia in refrigeration. However, we now know that this perceived environmental benefit is outweighed by the completely unanticipated impact of CFCs as ozone layer depleters, with great implications for human and ecosystem health.

The problem of CFCs in fact results from the very properties which commended their use. CFCs are chemically extremely stable, and consequently of very low toxicity. They are also volatile, hence their use as refrigerants. However, their chemical stability and non-toxicity meant that we felt little compunction in releasing them in large quantities into the atmosphere, where their volatility ensured long atmospheric residence times, allowing them to migrate to the stratosphere. There, through an extremely complex series of chemical reactions, CFCs are decomposed in the polar springs to release chlorine, which is the agent directly responsible for scavenging the protective ozone layer (Clarke 1992).

The problem of ozone depletion is different from that of global warming, in that measures to respond to ozone layer depletion could not be considered until the problem had (unexpectedly) occurred, been recognised and the cure diagnosed. In contrast, for global warming the uncertainty lies not in the recognition of the problem, but the understanding of what the implications may be.

Understanding and biodiversity loss

Biodiversity loss illustrates another form of environmental uncertainty; knowing that we are losing something of potentially great value, but not knowing what it is or what its value may be or have been. Loss of species has, necessarily, occurred since life began on Earth. Indeed, in the five billion years since the emergence of life, most of the species that have evolved have become extinct, with several periods of very rapid species loss perhaps being caused by asteroid impacts. Is it therefore a problem that species are becoming extinct now? Here the issue is not that species are becoming extinct, nor even that they are becoming extinct at a rapid rate. It is that we have no idea which species are becoming extinct, how many have been extinguished in the past century, nor of the ecological significance of the species lost. We do not even know how many species the Earth supports.

In summary, the example of the greenhouse effect, ozone layer depletion and biodiversity loss illustrate the problems of environmental knowledge as having the following aspects.

1   Even when we think we understand the basic principles of how human action

impinges upon nature, the ramifications of these effects may be so complex as to be beyond our powers of prediction (e.g. global warming).

2   We do not know the nature of the world, so that human action may have effects which we cannot predict (e.g. CFCs and ozone layer depletion).

3   The world is complex and heterogeneous, and many of its aspects may be damaged by human action without our knowing what that damage may be (e.g. species loss).

Understanding of human perceptions and motivations

As well as the problem of scientific knowledge, there is also that of our understanding of the processes of personal and social perceptions of environmental and social issues. We shall discuss three particular areas: the perceptions of 'needs', of 'nature' and of 'risk'.

Perceptions of needs

If we are to find effective policies to protect nature from human destructiveness, a first step is to identify the motives for this damaging production and consumption associated with our chosen lifestyles.

While modern economics seeks to understand the implications of human wants, it is rather poor at establishing their nature and source. Standard economic theory makes no significant distinction between goods which are luxuries and those which are basic needs, nor does it indicate how and why goods originally in the former category migrate to the latter (e.g. telephones and televisions in Western societies).

The standard utility theory analysis of neoclassical economics simply asserts that the consumption of goods and services gives the consumer well-being or 'utility', though at a diminishing rate as the consumption of that good or service increases. The implicit assumption in this approach is therefore that increased consumption possibilities increase individual happiness. Sadly, there is absolutely no empirical evidence to support this supposition.

Comprehensive surveys, in various countries over extended time periods, support two startling conclusions. First, the average self-assessment of well-being in a country is relatively constant over time; increasing average wealth and consumption have no effect on the trend figures. Second, within countries, those with higher consumption possibilities declare themselves to be happier than those with lower consumption possibilities, and this pattern is consistent across countries (Cantril 1965; Diener 1984: cited in Easterlin 1996). The conclusion is that while we may think that seeking to increase our consumption will make us happier, it will not. We only become happier if we increase consumption more rapidly than others in our society. Using Hirsch's (1977) term, happiness is a 'positional good'.

How, then, are we to understand human consumption and lifestyle choice? What motivates humans to seek expanded consumption possibilities? If we could formulate a distinction between needs, however defined, and luxuries, could this form the basis for rational considerations of consumption and lifestyles within environmental policy? Clearly, understanding of even this fundamental area is still poor.

Perceptions of nature

The very notion of nature, and how we perceive it, is contentious. Elsewhere (Proops 1989), one of us has discussed the range of perceptions of nature open to members of Western societies, and identified four 'paradigmatic images of the world'.

1. Undisturbed nature – the hunter–gatherer world

In this image of what constitutes nature and the natural, humans are present only as actors within nature, hunting and gathering much as do non-human animals. This image views modern humans as intrusive upon and destructive of nature. This seems to be a common image among North American environmental thinkers.

2. Humankind in nature – the agricultural world

This image sees humans as part of nature through harmonious agriculture and husbandry. The landscape is accepted as embodying humans and their works, but these are seen as humankind and nature in synergy rather than conflict. With a history of several thousand years of established agriculture in Europe, it is not surprising that this seems to be the dominant image among Western European environmental thinkers.

3. The human as creator – the industrial world

In this image, nature is the background to and inspiration for human achievements. Human resourcefulness and inventiveness rejoice in the challenge of nature. Nature is a tabula rasa upon which humankind can write its destiny. This image is close to some of the less mainstream branches of economic thought, such as the Austrian school (Rizzo 1979).

4. Gaia – the creative and self-sustaining world

Gaia (Lovelock 1979) is the world, all it contains, all it has been, all it might become. It created itself, and all elements within it work in harmonious ways to sustain it as it changes over eons. The destiny of individual species, or even whole groups of species, is unimportant. Gaia stands opposed to the other three images in that within it humans are not central, or even significant; in this it seems to express the views of some 'deep green' thinkers.

These are not offered as a comprehensive listing of attitudes to and images of nature, merely as a set which are reflected in the current environmental literature. Within the wider public, the concept of nature may carry yet further meanings and resonances, and work on social attitudes to nature is only just beginning in the academic literature (e.g. Dryzek 1997; Barry and Proops 1999). Until the nature and range of social understandings and perceptions of nature are clarified, environmental policy-making is working in a conceptual near-vacuum.

Perceptions of risk

Perhaps the most influential recent thinker on environmental risk has been Beck (1992), who argues that we now inhabit the 'risk society', where the risks of individual actions are borne more and more by society at large (e.g. nuclear risk). Another author on risk and its complexities is Adams (1995), who argues that risk is not something we passively bear and share. Rather, risk is an integral part of the way we structure our lives, and that when legislation reduces risk, we alter our behaviour and lifestyles to bring back our experience of risk to 'satisfying' levels; we have 'risk thermostats'.

At a more general level, the notion of 'risk' is difficult to apply to many environmental threats, as the understanding of risks by the 'science experts' is often enormously different from the understandings of the 'public'. The scientific assessment of risks to human health suggests that chemical plants are far more threatening than nuclear power stations. Over the past three decades, far more lives have been lost through accidents in chemical works than in the nuclear industry. However, the public perception of the risks associated with these two industries suggests that nuclear power stations are regarded as potentially extremely hazardous, while the risks associated with chemical plants is somehow regarded as more acceptable. This dissonance between statistical risk and perceived risk also applies to road versus air transport. Road transport carries a far higher mortality rate, both per hour travelled and per kilometre travelled, but public perception of the danger of air travel is focused on the rare cases of large scale mortality in (very rare) air crashes.

Clearly, not only is risk something which modern lifestyles makes more prevalent, the way we understand risk, and react to it, is extremely problematic, with major implications for the formulation and implementation of environmental policy.

Understanding of ethical systems and sustainability

Sustainability considerations are widely recognised as having both 'positive' and 'normative' aspects. The positive aspect concerns the achievement of outcomes which are, in some sense, sustainable, through the efficient application of natural science and social science understandings.

The normative aspect reflects the ethical dimension to human action, and suggests that both the nature of a 'sustainable world', towards which policy is striving, and the means by which this end is attained, are themselves open to assessments which lie outside the rather narrow range of 'efficiency' judgements. Rather, some actions towards humans, and the rest of nature, can be judged 'wrong', no matter how 'useful' such actions may seem in an instrumental sense.

This priority of ethical judgements over efficiency assessments is often asserted through statements such as: 'Ethics trump income'. That is, if an action is judged morally wrong then, no matter how great the material advantage it offers, that action should not be pursued. There is great resonance here with the current international negotiations over whaling. In recent years the debate has shifted sharply from considerations of the 'sustainable yield' of whales, to the notion that whales, as apparently intelligent mammals, should *not* be hunted.

Clearly, the issues of environmental knowledge and environmental ethics are not independent. As we discover more about natural systems, it is almost inevitable that what is considered 'morally considerable' in nature will alter. Conversely, understandings of the ethical positions held, explicitly or implicitly, within various cultures is itself an area where our knowledge is clearly incomplete.

The sources of environmental ethics

Theories of environmental ethics have developed over time from views which took the environment to be nothing more than a resource for human exploitation (*anthropocentric* ethics), through theories attaching moral considerability to individual animals and plants (*biocentric* ethics), to comprehensive environmental ethics which locate ethical considerability in every kind of natural living entity (*ecocentric* ethics).

Anthropocentric ethics

Historically, nature was conceived of as a God given body of resources to be manipulated to produce benefits for humans. Writers have attributed this anthropocentric view of nature variously to the Greek stoics (Passmore 1974), Christianity (White 1967), capitalism (Moncrief 1974) and technology (Moncrief 1974). Whatever its origins, there can be little doubt that this conception of nature has contributed greatly to the process of ecological devastation.

There have been attempts to modify the basic anthropocentric position in ways that give meaning to nature other than merely as an instrument for the satisfaction of human wants. One way to do this is to adopt a theological ethic of *stewardship* – a duty to God to care for nature (Montefiore 1970).

Fortunately, it is not necessary to assume a position of subservience to the Deity to reach a more environmentally benign position. Indeed, most intra-human ethics can be extended in ways beneficial to the environment. More enlightened versions of utilitarianism, for example, will have regard to the long-term human benefits of biodiversity (Wilson 1994) as well as the more immediate pleasures derivable from nature contemplation (Mill 1848). The anthropocentric basis for environmental protection is made more credible if future persons are taken into account (Sikora and Barry 1978).

Enlightened anthropocentric ethics all depend on consideration of human interests and are, therefore, subject to the same central criticism: the contingency of nature's value. If humans cease to regard nature, or some aspect of nature, as worthy of respect, or overlook its potential to fulfil human needs, then it becomes relegated to the status of inert material, 'resource', to be treated with moral indifference.

Biocentric ethics

One way to avoid the moral dependency of nature on human concern is to treat plants and animals as objects of moral concern *per se*. A leading example of this genre of environmental ethic is Regan's (1983) argument for animal rights. Regan asserts that all normal mammals of one year or more of age – a sub-category of animals which he refers to as 'subjects of a life' – have a moral right to life, since, he argues, subjects of a life are morally indistinguishable from mentally enfeebled human beings who are themselves generally accepted to possess such a right.

Although Regan's ethic is limited to a sub-category of sentient animals, more inclusive biocentric ethics have been developed. According to Van De Veer (1979) all organisms possess equal moral *considerability*, but not equal moral *weight*; the moral weight to be afforded to any given organism being a function of (a) the type of interest that is at stake and (b) the organism's psychological capacities. Taylor (1986) has developed a more egalitarian ethic based on *species egalitarianism* – the view that all species, including *Homo sapiens*, are morally equal. Taylor posits that all organisms are moral subjects because all organisms have 'inherent worth'. A sufficient and necessary condition for possession of inherent worth by organisms is that they have *a good of their own*; i.e. it is possible to imagine the world from their standpoint, to make judgements about what would be a good or bad thing to happen to them, and to treat them in such a way as to help or hinder them. For example, one can consider that although a tree does not feel pain when it is chopped down, it is nevertheless a bad thing to happen to the tree as it prevents the tree from fulfilling its potential life progression. Taylor maintains that to accept these premises leads one to adopt an attitude of *respect for nature* which, in turn, implies the acceptance of a number of specific duties towards all natural organisms.

Ecocentric ethics

Whilst biocentric ethics take us some considerable distance beyond the narrow anthropocentric position, it suffers from one weakness: it has no place for 'wholes' such as ecosystems. Ethics which focus on ecosystems are generally referred to as *ecocentric* ethics.

One of the problems of developing convincing ecocentric ethics is meeting criticisms of the concept of an 'ecosystem' that is the object of moral concern. If holistic concepts such as 'ecosystem' have no validity, no basis in reality, then there is no reason to incorporate them in an environmental ethic.

In 1935 the British ecologist Arthur Tansley published an influential article, '*The Uses and Abuses of Vegetational Concepts*' (Tansley 1935). Specifically, Tansley considered that the concept of a 'system' which carries no purposive or teleological connotations, would be better fitted to describe the relationships between organisms in the context of their whole physical environment. Thus the concept of the 'ecosystem' was born. Most ecologists now accept that ecosystems are the 'basic units of nature' (Evans 1956) and the most important concept in ecology (Cherett 1989). Likewise, in the main, environmental philosophers accept that ecosystems are real entities and, as such, appropriate objects of ethical concern.

One of the earliest ecocentric ethics is Leopold's land ethic. Leopold, game manager turned environmental philosopher, set out an ethic based on re-integrating humans into a *community* with the land (where 'land' is to be understood as 'soils, waters, plants, and animals . . . collectively') pithily summarised in two sentences: 'A thing is right when it tends to preserve the integrity, stability, and beauty of the biotic community. It is wrong when it tends otherwise' (Leopold 1966: 224–25).

In the land ethic, although both the individual organism and the biotic community are worthy of respect, the individual is ultimately subordinate to the community. The ethic demands not that *no* animals, plants or habitat ever be destroyed but that such destruction be avoided where it would damage the beauty, stability and diversity of the ecosystem as a whole (Callicott 1994).

Representative of more recent and philosophically sophisticated work is Rolston's (1988) combined biocentric and ecocentric ethic. Rolston's ethic is constructed on the premise that natural organisms and wholes possess 'intrinsic value'; i.e. value independent of their usefulness for human purposes.

In relation to individual organisms, Rolston ranks intrinsic value according to neural or phylogenetic complexity (humans above elks, elks above plants, plants above microbes and so on). Human individuals are, generally, ethically superior to non-human individuals, but human conduct towards other sentient animals should accord to a principle of non-addition of suffering and should avoid ecologically pointless suffering. Thus, hunting with a rifle for food is probably ethically justifiable, since the death pain is not greater than the suffering involved in natural deaths (but not if the animals are killed for fur coats since this is ecologically pointless).

In relation to non-sentient animals and plants Rolston proposes a principle *of non-loss of goods*. Thus using plants for human needs is permissible where the good that accrues to humans outweighs or equals the good lost in destruction of the plants. Eating plants for food or cutting timber to make housing would count as ethically permissible on this view.

Ethical conclusions

A number of conclusions can be drawn from the above overview of environmental ethics. The traditional narrow anthropocentric world-view is no longer an acceptable basis for human-nature relations, or for environmental law. Enlightened anthropocentric ethics have something to offer but are inevitably weakened by their tendency to arrogate human interests above those of the environment itself. Biocentric ethics take us a little further but these cannot provide proper justification for the preservation of wholes such as species and ecosystems. We need, therefore, to adopt a combination of biocentric and ecocentric ethics as the proper basis for environmental protection.

Understanding of the effectiveness of incentives and restrictions

If we are to attain sustainability for human and natural systems, we must implement appropriate policies, in large part through the institution of appropriate legal structures, both national and international. Clearly, such policies must be based on our understandings of the world, and also reflect our ethical positions, concerning humans (both in the present and the future) and the rest of nature (e.g. species extinction). We shall explore some of the implications of these joint requirements, and seek some guidance for policy and law, which would allow the target of sustainability to be achieved, while taking account of our limited, but changing, understandings of the world, and simultaneously respecting our ethical principles.

Environmental law and ethics

Analyses of 'environmental law' usually assume that a body of law specific to the environment can be identified as a discipline or bounded area of law in its own right. This in turn suggests that environmental problems can be resolved by the application of a discrete body of regulation to some fraction of human activities (generally assumed to be direct acts of pollution and direct acts of nature degradation). It is this flawed assumption that lies at the root of what we perceive as the weakness of environmental law. Ecocentric ethics stress the total inter-relatedness of all activities, states and substance. It is not the freakish, malicious, or extreme forms of human behaviour which threaten the environment, but the everyday lifestyles of the global population, especially those in the developed North.

Law which embodies ecocentric ethics, what can be called *ecological law*, would deal with the fundamental or institutional aspects of human-nature relations, and regulate areas of life usually considered to be outside law's proper domain. This would include (but not be limited to) challenges to many aspects of modern lifestyles; viz.:

•   unlimited mobility through private motor transport;

- use of large amounts of energy and the underlying reliance on fossil fuels;
- high consumption lifestyles;
- destruction of landscapes in the name of 'development';
- freedom of international trade; and
- reproduction of an unlimited number of children.

In this section we consider whether, were such a body of ecological law to emerge, it could retain its justification in liberal-democratic theory. This question is of importance because failure to achieve this, or some other jurisprudential justification, may delay the evolution of ecological law.

In addressing this question it is necessary, first, to distinguish the claims of liberalism from those of democracy. The central tenets of *liberalism* are freedom from state interference Mill (1972) and neutrality amongst conceptions of the good life (Sagoff 1988). Democracy's central claim, on the other hand, is that decisions should be made by the population, the *demos*, and not by some individual, monarch, committee or philosopher king. Liberalism does not necessarily imply democracy or vice versa. There have existed dictatorships, monarchies and forms of communism which have adhered to liberal norms, as well as democracies in which liberty has been greatly reduced by popular demand for security, order, etc.

Ecological law and liberal theory

The dominant view in modern Western jurisprudence is that law is, or should be, based on the tenets of liberalism and that law should not, therefore, interfere with matters of personal morality (Dworkin 1966). Non-liberals have responded that society has the right to enforce morality in order to ensure its own continued existence (Devlin 1968); that morals must be enforced for their own sake (Stephen 1967); that the law must draw a line between what is acceptable in a civilised society and what is not; that law cannot, in fact, remain morally neutral (Mitchell 1967); and that law which fails to reflect a growing body of moral opinion will lead to the population taking matters into their own hands (Dworkin 1966). The latter reason is of considerable importance as environmental activists and 'eco-warriors', dissatisfied with the paltry, reactionary and piecemeal environmental law project, have shown that they are prepared to act on their moral beliefs and public sense of outrage by taking the law into their own hands (Foreman 1991; List 1993).

In his examination of the relationship between environmentalism and liberalism Sagoff has concluded that environmentalists can be liberals, implying that environmental law is consistent with, and justified by, the tenets of liberalism. The key to Sagoff's conclusion is his assumption that few if any existing environmental laws transgress liberalism's basic demand for a core of private behaviour and beliefs (Sagoff 1988: 183):

Whilst it is possible that *environmental* law may be compatible with this weak form of liberalism, there are a number of reasons why *ecological* law is not. First, ecological law makes nonsense of liberalism's proscription of interference except on grounds of harm to others. Ecocentric ethics, as we have seen, demonstrate that plants, animals, species and ecosystems are morally considerable bearers of intrinsic value. These entities must, therefore, assume the status of 'individuals' in determinations of the limits of legal or state intervention. Since all of nature consists of organismic or communitarian 'individuals', and since *all* human activity has some detrimental environmental consequences, liberalism's claim to rope off a safe area of private 'law free' behaviour is misguided.

There is a further, perhaps more profound, reason why ecological law cannot draw on liberal theory for justificatory support. As critics have often observed, liberalism is underlain by the ideology of *individualism*: a model of the human psyche which is in considerable tension, not to say conflict, with environmental ethics that take groups or wholes as their subject matter. Communitarians reject this conception of the human condition; maintaining that humans are not self-creating uninfluenced selves (MacIntyre 1985) and point out that ethical life requires adoption of *social* roles. Recovery from alienation, they point out, requires a reversal of the decay in the locatedness that individuals once received from the institutional fabric: a locatedness lost through technology, industrialisation, bureaucracy, urbanisation, population growth, and by the treatment of nature as *other.* Ecological law must, therefore, be built on *public* recognition of ecocentric and biocentric values. Environmental harm is a matter for the community requiring communal decision-making, not 'individual consumers expressing personal wants' (Freyfogle 1994: 843).

If ecological law is inconsistent with classical liberal theory then other, non-liberal, theories will have to provide its jurisprudential underpinnings. 'Natural law' theories appear, appropriately, to hold considerable promise in this matter. The notion that human affairs are properly ordered when they accord to an underlying and pre-existing 'natural' order is one which appeals to many environmentalists. On this view we should order our affairs with nature according to 'ecological practicable reasonableness' (after Finnis 1980). On this view the preservation of the planetary life-support system, the reduction of the global human population, the strict control of human eco-destructive activities, the legal embodiment of inherent ecological values, would all properly be considered to be goals which any coherent and effective body of law must strive to achieve. Natural law jurisprudence would imply that laws should 'be chosen according *to an ideal goal or good,* rather than represent the haphazard implementation of voters' preferences, be they popular, or those of specific interest groups' (Westra 1993). In this jurisprudence we would expect to see an increase in judicial activism (Dias 1994) and the development of new legal principles to guide the evolution of the law in novel matters and statutory interpretation.

Ecological law and democratic theory

The desirability of democratic participation in the formulation and enforcement of environmental law is usually taken as a given by environmentalists. However, we should not assume that democracy is implicit in, and can be derived from, environmental ethics: this would be to confuse theories of value (about nature) with theories of agency (how to act politically) (Goodin 1992). Ethics are determined not by what is popular, but by what is good, right or virtuous. Popular support, especially local support, for policies that are environmentally destructive but which provide tangible short-term benefits is to be expected in a world of self-interested individuals. Democracy cannot, therefore, be identified as a green 'good' to be ranked alongside other green values.

The tendency of democracy to favour the selection of short-term environmentally destructive policies may be countered by the entrenchment of 'pre-democratic' rights (Harrison 1993) and education (Westra 1993: 128); providing, respectively, barriers against the selection of immoral policies, and encouragement towards the common good through enlightened leadership. The first of these – pre-democratic entrenchment of ecological values – may be achieved either by adopting an environmental Bill of Rights and/or by embedding ecological values in constitutional law. Either would elevate ecological law to take precedence over other legal and policy choices.

Ecological law and education

The second technique, education, is essential to the initial acceptability of ecological law. 'Public education programmes may be required both before and after the law comes into force to help the public understand and support it' (IUCN, 1980: para. 11.9). Education is, in this context, to be used not only as information *about* environmental ethics but also to *induce* the production of those values in a wide proportion of the population. We may expect that, in the future, education in ecological values will be included in mainstream education programmes. Education in environmental ethics will be especially important for all of those who may hold positions of political power since, both in theory and practice, legislators behave as intellectually independent representatives, not mere delegates (Mill 1912) often giving the population only options of form, not substance, on key issues (Plamenatz 1973). Empirical evidence encouragingly suggests that government officials who deal with the negotiation of environmental instruments often hold quite deep personal ecological values (Craig and Glasser 1993).

In attempting to assess the likelihood of widespread and successful education in ecological values one should not overlook ecological law's own educative function. This derives both from increasing dissemination and knowledge of the substantive content of environmental law and from increasing opportunities to become actively involved in the machinery of environmental law. Psychological theories of moral development suggest that knowledge of laws, and acting from a sense of legal obligation, are important intermediate stages in the path to higher stages of personal ethical development (Kohlberg 1984). Just as the knowledge that murder is *illegal* and severely punishable adds to our moral understanding that murder is *wrong*, so too knowledge of the content of ecological law would

contribute to the development of personal ethical values that favour environmental protection. Furthermore, recent 'narrative' studies of moral development
stress the importance of contextual and cultural influences, especially language, in
the development of moral thinking (Tappan 1997). So, just as the ordinary meanings of so-called legal terms (e.g. 'infant', 'corporation') influence their judicial
interpretation (Stone 1972: 488), so too the language of law ('legal right', 'trust',
'duty of care', 'offence', etc.) forms part of the narrative of right living in individuals. Ecological law would, in proclaiming the 'rights' of nature and duties to
avoid nature destruction, act to transform personal values.

Conclusions

In this chapter we have attempted to engage with the issue of sustainability, and
sustainable lifestyles, by considering problems of knowledge, ethics and law that
confront policy makers. Our tentative conclusions are as follows:

1   Knowledge is a fundamental issue for environmental policy. In particular, our
    understanding of how human lifestyles and actions impinge upon nature is
    beset with difficulties. Further complexities arise when we seek to understand
    why certain behaviours and lifestyles are preferred, and why there are such
    divergent attitudes towards nature, and towards the risks which our lifestyles
    entail.
2   There are various ways in which ethics can seek to embrace concerns about
    nature, and the damage our lifestyles impose upon it. While an ecocentric
    approach is, we believe, both the most justified and most effective attitude,
    there is continuing debate on the alternative merits of anthropocentric and
    biocentric ethics.
3   An ecocentric ethic should be reflected in ecological law, though the consist-
    ency of such a legal framework with either standard liberal or democratic
    theories. Ecological law would require greater sacrifices of personal liberty
    than liberal theory would allow. However, while ecological law would also
    challenge democratic theory, it could be supportive of such principles as an
    educator and originator of personal ethical values.

References

Adams, J. (1995) *Risk*. UCL Press, London.
Barry, J., and Proops, J.L.R. (1999) Seeking sustainability discourses with Q method-
    ology. *Ecological Economics* (forthcoming).
Beck, U. (1992) *Risk Society: Towards a New Modernity*. Sage, London.
Callicott, J.B. (1994) The conceptual foundations of the land ethic. In: C. Pierce and
    D. Van De Veer (eds), *People, Penguins and Plastic Trees*, 2nd edn, Wadsworth,
    Belmont.
Cantril, H. (1965) *The Pattern of Human Concerns*. Rutgers University Press, New
    Brunswick, NJ.

Cherrett, J.M. (1989) Key concepts: the results of a survey of our members' opinions. In: J.M. Cherrett (ed.) *Ecological Concepts*, Blackwell, Oxford.

Clarke, A.G. (1992) The atmosphere. In: R. Harrison (ed.) *Understanding Our Environment*, Royal Society of Chemistry, Cambridge.

Craig, P.P., and Glasser, H. (1993) Ethics and values in environmental policy: the said and the UNCED. *Environmental Values* 2:137–57.

Devlin, P. (1968) *The Enforcement of Morals*. Oxford University Press, Oxford.

Dias, A. (1994) Judicial activism in the development and enforcement of environmental law. *Journal of Environmental Law* 6:243–62.

Diener, E. (1984) Subjective well-being. *Psychological Bulletin* 95:542–75.

Dryzek, J.S. (1997) The Politics of the Earth: Environmental Discourses. Oxford University Press, Oxford.

Dworkin, R.M. (1966) Lord Devlin and the enforcement of morals. *Yale Law Journal* 75:986–1005.

Easterlin, R.A. (1996) *Growth Triumphant*. The University of Michigan Press, Ann Arbor.

Evans, F.C. (1956) Ecosystems as the basic unit in ecology. *Science* 123:1127–28.

Faber, M., Manstetten, R., and Proops, J.L.R. (1996) *Ecological Economics: Concepts and Methods*. Edward Elgar, Cheltenham.

Finnis, J.M. (1980) *Natural Law and Natural Rights*. Clarendon Press, Oxford.

Foreman, D. (1991) *Confessions of an Eco-Warrior*. Harmony Books, Boston.

Freyfogle, E.T. (1994) The ethical strands of environmental law. *University of Illinois Law Review* 819–846.

Goodin, R.E. (1992) *Green Political Theory*. Polity Press, Cambridge.

Harrison, R. (1993) *Democracy*. Routledge, London.

Hirsch, F. (1977) *Social Limits to Growth*. Routledge, London.

Houghton, J., Meira Filho, L., Callander, B., Harris, N., Kattenburg, A., and Maskell, K. (eds) (1996) *Climate Change 1995: The Science of Climate Change, contribution of Working Group I to the Second Assessment Report of the Intergovernmental Panel on Climate Change (IPCC)*. Cambridge University Press, Cambridge.

IUCN (International Union for the Conservation of Nature and Natural Resources) (1980) *World Conservation Strategy: Living Resource Conservation for Sustainable Development*. Gland, Switzerland.

Kohlberg, L. (1984) *The Psychology of Moral Development*. Harper and Row, New York.

Leggett, J. (ed.) (1990) *Global Warming*. Oxford University Press, Oxford.

List, P. (1993) *Radical Environmentalism: Philosophy and Tactics*. Wadsworth, Belmont.

Leopold, A. (1966) *A Sand County Almanac*, Oxford University Press, New York.

Lovelock, J. (1979) *Gaia: A New Look at Life on Earth*. Oxford University Press, Oxford.

MacIntyre, A. (1985) *After Virtue: A Study in Moral Theory*. University of Notre Dame Press, Notre Dame.

Mill, J.S. (1848) *Principles of Political Economy*, London.

Mill, J.S. (1912) *Considerations on Representative Government*. Oxford University Press, Oxford.

Mill, J.S. (1972) On liberty. In: H.B. Acton (ed.), *Utilitarianism, On Liberty, Considerations of Representative Government*, Dent, London.

Mitchell, B. (1967) *Law, Morality and Religion in a Secular State*. Oxford University Press, London.

Moncrief, L. (1974) The cultural basis for our environmental crisis. *Science* 170: 508–12.

Montefiore, H. (1970) *Can Man Survive?* Fontana, London.

Passmore, J. (1974) *Man's Responsibility for Nature*. Scribners, New York.

Plamenatz, J.P. (1973) *Democracy and Illusion*. Longman, London.

Proops, J.L.R. (1989) Ecological economics: rationale and problem areas. *Ecological Economics* 1:59–76.

Regan, T. (1983) *The Case for Animal Rights*. Routledge, London.

Rizzo, M.M. (1979) *Time, Uncertainty and Disequilibrium*. Heath, Lexington, Mass.

Rolston, H. (1988) *Environmental Ethics: Duties to and Values in the Natural World*. Temple University Press, Philadelphia.

Sagoff, M. (1988) *The Economy of the Earth: Philosophy, Law and Economics*. Cambridge University Press, Cambridge.

Sikora, R.I., and Barry, B. (1978) *Obligations to Future Generations*. Temple University Press, Philadelphia.

Stephen, J.F. (1967) (originally 1874) *Liberty, Equality, Fraternity*. Cambridge University Press, Cambridge.

Stone, C. (1972) Should trees have standing? Toward legal rights for natural objects. *Southern California Law Review* 45:450–501.

Tansley, A (1935) The use and abuse of vegetational concepts and terms. *Ecology* 16:284–307.

Tappan, M.B. (1997) Language, culture and moral development. *Developmental Review* 17:78–100.

Taylor, P.W. (1986) *Respect for Nature: A Theory of Environmental Ethics*. Princeton University Press, Oxford.

Van De Veer, D. (1979) Interspecific justice. *Inquiry* 22:55–79.

Von Weizsäcker, E., Lovins, A. and Lovins, H. (1997) *Factor Four: Doubling Wealth, Halving Resource Use*. Earthscan, London.

Westra, L (1993) The ethics of environmental holism and the democratic state: are they in conflict?' *Environmental Values* 2:125–36.

White, L. Jr. (1967) The historical roots of our ecological crisis. *Science* 155:1203–7.

Wilson, E.O. (1994) *The Diversity of Life*. Penguin, London.

3   Environmental policy-making

What have economic analysis and the idea of sustainability got to offer?

Paul Ekins

Introduction

Concerns about 'the environment' cover an enormous range of issues, from the global problems of climate change, ozone depletion and the loss of biodiversity, to more localised issues of air and water quality, to parochial concerns such as litter, the provision of and access to urban green space and the existence of local environmental features such as trees or the village pond. The purpose of this part of the book is to explore how public policy should seek to address these issues. The specific focus of this chapter is the role of both economic analysis and the concept of sustainability in forming and informing environmental policy.

At the most general level, there are two principal concerns about human societies' current uses of the environment. The first is that these uses are unacceptably damaging people's quality of life in the present (for example, in different countries, through air pollution, massive 'development' projects or the destruction of valued landscapes). A complicating factor in this situation is that the victims of the damage are very often not the same people as have benefited from the damaging activity. The second concern is that the environmental impacts from these uses may be such as to seriously reduce the quality of life, and perhaps even the prospects for survival (for example, through climate change or biodiversity loss) of people in the future. Environmental concerns, therefore, inevitably involve considerations of fairness within the human generations currently alive, and between present and future human generations. For those who believe that other species have intrinsic value, there is also the issue of what balance should be struck between human welfare and the rights to existence of these other species.

Economics, in general, has little guidance to give about issues of intragenerational and intergenerational equity. Indeed, the former is often regarded as something to be traded off against economic efficiency. With regard to the latter, the allocation of resources between the present and the future, this is usually mediated through the discount rate, which is normally assumed to be positive (so that future costs and benefits are worth less than those in the present) on the uncertain grounds that future generations are likely to be richer than those currently alive. As far as environmental assets are concerned, on current trends this is most unlikely to be true. On the intrinsic value of other species, economics, as a quint-essentially anthropocentric mode of analysis, is necessarily silent. It is clear that, whatever economics has to offer environmental policy, the distributional dimension of human environmental impacts means that it cannot address environmental problems on its own.

Leaving issues of equity and distribution to one side, it might be thought that economics was well equipped to analyse the purely allocative dimension of uses of the environment. The role of economic analysis with regard to the environment, and how this role has developed over time, is the subject of the next section.

Economics and the environment

Early history

The obvious importance of natural resources and the environment to economic activity and to human life in general has caused economists to be concerned about these issues since the dawn of economics as a mode of analysis. Two of the earliest classical economists – Thomas Malthus and David Ricardo – were much exercised by the prospect of a growing human population in the context of a fixed quantity of agricultural land of differing fertility.

Malthus' theory of population, for which Ricardo expressed his admiration (Ricardo 1973: 272) contended that: 'The power of population is indefinitely greater than the power in the earth to produce substance for man. Population, when unchecked, increases in a geometrical ratio. Subsistence increases only in an arithmetical ratio. . . . This implies a strong and constantly operating check on population from subsistence' (Malthus 1970: 71). Should somehow more food become available, then 'population does invariably increase' (Malthus 1970: 79), maintaining the mass of the population at the basic level of subsistence.

While there were differences between Ricardo's and Malthus' analyses of the economic consequences of these tendencies, both perceived that population could only grow by taking more land into production which, assuming that the most fertile land had been used first, would be of diminishing fertility. Taking this land into production would increase the price of food and therefore raise the subsistence wage, reducing the profitability both of the marginal agriculture and of industry. This erosion of profitability in both agriculture and industry reduces the incentive to invest, with profits 'sinking to the lowest state required to maintain the actual capital' (Malthus 1974: 282), resulting at length in a 'stationary state' of the economy with stagnant profits and investment and population growth constantly kept in check by starvation. This was the vision that later led Thomas Carlyle to refer to political economy as the 'Dismal Science'.

Both Malthus and Ricardo recognised that the onset of the stationary state could be postponed by technological progress or a halt in population growth before the stationary state was reached. Technological progress could result in either an increase in the productivity of land or in the substitution of capital for labour, thereby increasing labour productivity. Both effects would counteract the tendency for profits to fall and maintain the incentive to invest. Contrary to his 'dismal' image, Malthus even perceived that improvements in agriculture 'may certainly be such as, for a considerable period not only to prevent profits falling, but to allow for a rise' (Malthus 1974: 282).

In retrospect, Malthus' hopes for technological progress have proved justified. Technological advance has permitted more people to live at higher material standards than Malthus ever conceived. His perception of the overwhelming power of population growth has been contradicted by those industrial countries which combine high living standards with more or less zero population growth. Future population growth, of course, may yet engender Malthus' dismal scenario, though such a scenario is more likely to come to pass through social and political upheaval due to overpopulation, rather than a quietistic submission to natural forces. What the years since Malthus have shown incontrovertibly is that high living standards do not engender population growth; that population growth can be reduced and halted; and that the productive power of technology is enormous. It is as clear today as it was to Malthus, that the stationary state is not an inevitability.

The exhaustibility of resources

With the exception of solar radiation, which enters the biosphere in a quantity and with a geographic distribution that is beyond human influence, all environmental resources that are used for economic activity are exhaustible. Some resources, such as concentrated mineral ores and fossil fuels, are non-renewable and are necessarily depleted with extraction. Others are conditionally renewable, dependent on rates of harvest and ambient environmental quality.

The economics of resources has traditionally focused on the optimal depletion of the resources, where 'optimal' refers to the maximisation of the profit to the resource's owner from its sale, or of the social welfare to be derived from its consumption. At the optimal rate of depletion, the stock of the resource and the demand for it will fall to zero at the same moment.

From the private owner's point of view, the key economic insight is that, in a perfectly functioning market, the rate of depletion would be such that the price of the resource net of its extraction costs (i.e. its price 'in the ground', also called its rent) would rise at the same rate as the rate of interest (Hotelling 1931). The intuition behind 'Hotelling's Rule' is easily seen. Were the resource's rent to be rising more slowly than the rate of interest, its price would fall, as owners sought to acquire alternative assets instead. This would tend to restore the resource's rate of rent increase to parity with the interest rate. Were the rent to be rising faster than the rate of interest, the opposite would happen: people would seek to buy stocks of the resource, the net price of which would rise, slowing its rate of rent increase, again until it was equal to the interest rate.

Theoretical results for perfect markets have been modified to try to take into account market imperfections such as non-exclusive ownership (e.g. an oil-field that extends under differently owned plots and is tapped into by differently owned wells) or varying degrees of monopoly. As would be expected, the former condition leads to more rapid depletion than the optimal rate, as each owner tries to maximize their private return; the latter condition leads to slower depletion than the optimal rate, due to the restriction of output that the concentration of ownership brings about. Dasgupta and Heal (1979) is the classic work that reviews and derives such results, the mathematics of which can be formidably complex.

An empirical evaluation of Hotelling's Rule (Smith 1981) does not support its practical validity. Smith examined the price movements of twelve non-renewable resources: four fossil fuels and eight metals, through the use of five different economic models, including the simple Hotelling model. For four of the metals, none of the models used had any explanatory power. Hotelling's Rule was only accepted by the data for two of the resources. While the best-performing model, Heal and Barrow (1980), was accepted by eight of the resources, in all but three cases even it was outperformed by a simple autoregressive model that related the resource's current price to that in the previous time period. Smith's conclusion is that variables not entering the models, such as extraction costs, new discoveries, and changes in market structure and their institutional environment, for which data is not generally available in suitable form, must also be important in explaining price movements.

Where privately optimal depletion seeks to maximize the present value of profits from a resource over its lifetime, socially optimal depletion is concerned with maximizing the social utility to be derived from it. Utility is normally identified with consumption, so the problem then becomes one of maximizing the present value of the consumption of the resource through time. Several conditions affect the optimal depletion path: whether the resource is essential to the production of the consumption good; whether, and to what extent, forms of produced capital can be substituted for the resource; whether technological change either economises on the use of the resource or develops a substitute that renders it inessential; whether a new resource will be discovered that serves the same purpose; and the size of the discount rate, the relative value that is given to present and future consumption.

The core theoretical result in this area is that if the discount rate is positive (i.e. if future consumption is worth less than present consumption), if the resource is essential to consumption, and no technological breakthrough, discovery of substitutable resources or substitutability with produced capital stop it from being so, then it will be optimal to deplete the resource fully and, therefore, drive future consumption to zero. Improvements in the efficiency of resource use (the yield of more consumption goods per unit of resource use), or a reduction in the discount rate, will prolong consumption but will not prevent its eventual decline. This can only be achieved by rendering the resource inessential for consumption by the development or discovery of substitutes. In this case, the resource may still be fully depleted but, as far as the maintenance of consumption is concerned, it will not matter.

The difference between a renewable and non-renewable resource is the capacity of the former for self-regeneration. Provided the harvest-rate does not exceed the rate of regeneration, the stock of the resource will be undepleted. Pearce and Turner (1990: 272) derive a 'basic equation for optimal renewable resource use', which yields the following principal conclusions about renewable resources:

A   Clearly defined property rights and profit maximisation, with a zero discount rate, will tend to result in a yield that is lower and a stock that is higher than that at maximum sustainable yield.

B   Open access conditions will reduce the stock from the level in A, but will not result in extinction provided that the sustainable yield is not exceeded.

C   A positive discount rate will tend to result in a stock level between that of A and B. The higher the discount rate the closer will the stock level be to B.

D   If the discount rate exceeds the net rate of return from the resource as an asset, the resource will be liquidated, perhaps to extinction.

E   Increasing the price of the resource, or reducing the cost of harvesting it, will reduce the stock level. If the price is above the cost at low population levels, then extinction becomes likely.

F   The calculated costs of harvesting the resource should include the externalities of both attendant environmental damage costs and foregone option and existence values. They frequently do not do so, increasing harvesting beyond the social optimum.

Exhaustion and extinction: the evidence

Malthus and Ricardo were concerned that population levels would outstrip either the available quantity or the fertility of agricultural land, but the first economist who worried about the actual depletion of resources was W. S. Jevons, whose book *The Coal Question* warned of the dire consequences for British industry of, in his view, the inevitable exhaustion of British coal stocks and consequent increases in coal prices. 'The exhaustion of our mines will be marked *pari passu* by a rising cost or value of coal; and when the price has risen to a certain amount comparatively to other countries, our main branches of trade will be doomed' (Jevons 1965: 79).

In the event, Jevons has been proved to have underestimated, as have many conservationists since, the ability of human ingenuity under the influence of market forces to discover new reserves, develop substitutes and increase the efficiency of use of potentially scarce resources. However, this does not guarantee that human ingenuity will always overcome scarcity, the evaluation of which remains an empirical matter, over which expert opinion is divided.

Part of the reason for this division is the use of different definitions and indicators of scarcity. Barnett and Morse's classic 1963 investigation, later updated by Barnett (1979), using a unit cost indicator, found no increase in scarcity since the late nineteenth century over a wide range of resources, with the exception of forest products. Hall and Hall (1984) conversely found that coal increased in scarcity on a unit cost test, but on a relative price test oil, gas, electricity and timber all exhibited scarcity increases through the 1970s.

Most economists have followed Barnett and Morse's conclusion that natural resources, at least those that are priced inputs into production, are unlikely to become scarce enough in the foreseeable future to act as a constraint on production. The best known expression of the opposite view, *Limits to Growth* (Meadows *et al.* 1972), updated two decades later (Meadows *et al.* 1992), warns of a significant danger of industrial collapse on current trends, due to the combined effects of population growth, resource depletion and pollution, by the middle of the twenty-first century. The two opposing views derive entirely from the different assumptions they make about new technological developments that promote substitution, resource efficiency and structural economic change.

There is far less controversy over the empirical evidence for extinction, which, despite uncertainty over overall numbers, is generally agreed to be proceeding at a rate unprecedented since the extinction of the dinosaurs 65 million years ago. Some estimates suggest that 10–20 per cent of a possible 10 million species (though the number could be as high as 30 million) will have disappeared by the year 2000.

There are several reasons why this matters economically even if no intrinsic value is given to other species. Many people enjoy watching wildlife, or even just knowing that it exists. Plants have been the source of valuable medicinal drugs and as yet undiscovered species which may be extinguished could doubtless yield more. Genetic diversity is important in agricultural breeding programmes. Perhaps most importantly, complex ecosystems, such as rainforests, provide natural services, such as climate regulation, on which many people are dependent. Ignorance about the functions of such ecosystems is still very great. Their destruction, which is irreversible, could prove catastrophic.

The economics of renewable resources discussed in the last section gives a clear understanding as to why extinction is occurring: prices of resources generally do not reflect ecosystem functions or option or existence values; costs of harvesting can be low; discount rates can be above the sum of natural growth rates and rates of appreciation of the natural stock; open access conditions often pertain; and livelihoods are often dependent on continuing unsustainable exploitation.

Environmental degradation

With the increasing scale of industrial activity in the nineteenth century, the effects in terms of environmental degradation soon became obvious. John Stuart Mill was the first economist to recognise that the growth of production might be at the expense of environmental enjoyments:

> It is not good for man to be kept perforce at all times in the presence of his species. A world from which solitude is extirpated is a very poor ideal. . . . Nor is there much satisfaction in contemplating the world with nothing left to the spontaneous activity of nature. . . . If the earth must lose that great

portion of its pleasantness which it owes to things that the unlimited increase of wealth and population would extirpate from it, for the mere purpose of enabling it to support a larger, but not a better or happier population, I sincerely hope, for the sake of posterity, that they will be content to be stationary, long before necessity compels them to it.

(Mill 1904: 454)

The unintended and uncompensated loss to one person of natural beauty, pleasantness and solitude in nature due to the economic activity of another is an example of what, following the analysis of A. C. Pigou, has come to be called an 'externality'. Another more often quoted example is pollution through the discharge of wastes from production or from the disposal of the products themselves.

Although Pigou never used the term 'externality', his description of the effect remains definitive to this day: 'The essence of the matter is that one person A, in the course of rendering some service, for which payment is made, to a second person B, incidentally also renders services or disservices to other persons (not producers of like services), of such a sort that payment cannot be exacted from the benefited parties or compensation enforced on behalf of the injured parties' (Pigou 1932: 183).

Environmental externalities arise because property rights to the use of environmental resources are either non-existent – the resources are treated as 'free goods' – or ill-defined. In principle, as suggested in a celebrated paper by Ronald Coase (1960), the externality problem may be solved by the clear legal delineation of these rights, so that environmental conflicts may be resolved through private negotiation. In practice, it may not be feasible for political or other reasons to give strict definition to property rights over natural resources. It is not clear, for instance, what system of private ownership could realistically encompass the atmosphere or the stratospheric ozone layer. Alternatively, it may be that, even if such resources could be privately owned, their degradation would affect so many people that the transaction costs involved in negotiations would be so great as to prohibit the negotiations taking place. Consequently the property rights approach to the resolution of the problem of externalities, while theoretically appealing, is often practically infeasible.

Another theoretically attractive approach to resolving the externality problem is to seek to 'internalise' the cost by levying a charge or tax on the activity concerned, or on the environmentally damaging effect. This causes producers to reduce the amount of damage they cause. Theoretically, it is possible to arrive at a point of 'optimal' environmental damage at which the marginal cost of the damage is precisely equal to the marginal benefit of the activity that causes it. However, this requires complete knowledge of the economic value of the damage at different levels of the activity. The next section explores how such values are derived.

Environmental accounting and valuation

It is clear that environmental resources have value in production and consumption and can in principle be dealt with like other production inputs and consumption goods and services. It is also clear that the environment performs valuable services in providing a depository for the wastes of production and consumption and in providing services of amenity and recreation. Again, in principle, these services can be valued. However, the environmental functions that contribute to human life, and the environmental damage caused by economic activity, often occur outside of market exchanges and are unpriced. One result of this is that they have often been excluded from economic consideration.

An example of this exclusion is that environmental changes have tended not to feature in the national accounts, which measure the flows of goods and services generated by the economy. Yet it is clear, on the one hand, that environmental resources are the prerequisite of much economic activity, and, on the other, that the loss or gain in environmental quality as a result of economic activity should be subtracted from or added to the monetary value deriving from that activity. In the 1980s there were concerted attempts to bring environmental values within the remit of national accounting, culminating in 1993 in a new United Nations Handbook on Integrated Environmental and Economic Accounting (UNSD 1993). Because of the difficulties in arriving at credible estimates of the costs of environmental damage, for the reasons given below, other methods of valuing environmental effects have been used in this work. Broadly speaking, resource depletion is valued at the user cost (resource price less extraction costs) of the resource; and environmental degradation is valued at the cost of restoring the environment to its pre-degradation level. The techniques and detailed comments on them are given in Lutz (1993).

Now that environmental issues have risen in perceived importance, and in order to accommodate them into normal methods of economic appraisal, there have recently been intensive efforts to give money values to the environmental functions that contribute to human life, and to the environmental damage that arises as a result of human activities. The valuation of unpriced environmental functions or damage uses techniques of two kinds: indirect market methods, which seek to infer environmental values from market choices for other goods; and direct methods which survey people's willingness to pay for the environment.

The indirect methods can be divided into three kinds (Cropper and Oates 1992: 703ff.): those which derive from behaviour which seeks to avert or mitigate the damage caused; those which exploit the complementarity of a purchased good with the relevant environmental good; and 'hedonic market' methods. An example of the first kind is the purchase of a medicament against the effects of air pollution; an example of the second is the cost of travel to a site of environmental quality; an example of the third is the increase (or decrease) in the price of a house which benefits (or suffers) from an environmental amenity, such as a beautiful view (or disamenity, such as aircraft noise). In each case the environmental good or bad is valued at the cost of purchase of all or part of some other good.

Direct methods, sometimes called contingent valuation methods (CVM) use the direct questioning of people to establish their willingness to pay for environmental goods, or to avoid environmental bads, or to accept compensation to forgo the former or suffer the latter. The outcome of such surveys is obviously greatly influenced by the design of the questions and the amount of information about the relevant issue which the respondent either possesses or is given. Even with careful attention to these and other issues, however, the method has been criticised on such grounds as that: there is a difference between professing a willingness to pay and actually paying; people may resent being asked to pay for certain environmental goods and this may influence their answers; or they may seek to 'free ride' on others' professed payments. The evidence also shows that elicited values differ depending on the number of issues that are included in any single survey. Such considerations cast doubt on the appropriateness of comparing values derived from CVM with actual market values, and caused Diamond and Hausman (1994: 46, 62) to conclude that CVM 'is a deeply flawed methodology for measuring non-use values' and that reliance on it 'in either damage assessment or in government decision-making is basically misguided'.

The problem is that CVM is the only method that has yet been devised of making any kind of estimate of certain crucial environmental values. Pearce (1993: 16ff.) has divided such values into use, option and existence values, where the first is the value of use now, the second is the value of preserving the option for use in the future, and the third is the value of simply knowing that something exists. Only CVM can attempt to estimate existence values and those option values for which no well-functioning futures market exists.

It is clear that the most basic conditions of existence, without which production and consumption, and human life, would not be possible at all are provided by the carbon, nitrogen, water and other biogeophysical cycles, working through the biological diversity of ecosystems. It is here, even more than with the resources and waste-absorption services of the environment, that the 'in principle' possibility of valuing the environmental contribution to human welfare is thwarted by the inability in practice of the various techniques of economic valuation to generate credible money values, even to within an order of magnitude, of the worth of some of the most important environmental functions.

The controversies over environmental valuation are not surprising given the common characteristics of contemporary environmental concerns, which make them most intractable for economic analysis and which include chronic uncertainty often verging on the indeterminate; irreversibility; profound social and cultural implications; actual or potential grave damage to human health, including threats to life; global scope; and a long-term intergenerational time-scale.

There is rarely any generally acceptable way of putting a money value on costs with these characteristics, especially when the characteristics are combined. CVM and other techniques of environmental valuation are not able realistically to assess the costs of displacing millions of people from low-lying coastal areas (global warming); of hundreds of thousands of extra eye-cataracts and skin cancers (ozone depletion); of other processes of large-scale environmental degradation, such as current rates of deforestation, desertification and water depletion, which entail considerable national or international threats to life and livelihood; of the possible unravelling of ecosystems (species extinction); of the persistent release of serious toxins (e.g. radiation) or the effects of major disasters (e.g. Chernobyl, Bhopal).

If a large fraction of the benefits to be derived from the environment, and hence the costs of environmental damage, defy monetary valuation, then economics cannot identify the situation of 'optimal use' of the environment, that ideal point of marginal balance at which the loss of one unit of environmental quality or resource precisely equals the benefit to be derived from its sacrifice. Indeed, because of the incommensurability of environmental and economic values, not only can the 'optimal use' situation not be identified. It cannot even be said to exist.

Environmental policy cannot, therefore, be driven by the quest for optimisation. If it is not simply to be determined on an ad hoc basis by the strongest among conflicting interests in any given situation, it needs another guiding principle. Environmental benefits and costs need to be considered in a different decision-making framework. It is here that the concept of sustainability has a potentially important role to play.

Making use of the environment sustainable

Principles and standards of environmental sustainability

As noted in the Introduction to this book, the use of the term 'sustainability' with reference to development and the environment has been plagued by imprecision and confusion. Yet the word itself has a well understood meaning: the capacity for continuance into the long-term future. An activity or situation is sustainable if it can be projected to continue indefinitely into the future; conversely, one which cannot is unsustainable.

Sustainability is an important and familiar concept in economics. 'Sustainable' as an adjective is most often applied to income, or consumption, in a situation where the capital stock from which the income was derived has been maintained. In fact, income is actually strictly defined in economics as the amount that is left for consumption in any given period *after* any depletion or depreciation of the capital stock has been taken into account. Unfortunately it is only recently that the environment has started to be considered as part of the wealth-creating capital stock.

It is the human use of the natural environment that has given rise to current concerns about sustainability. It is feared that industrial economies, societies and ways of life are not sustainable because they are treating the natural environment in such a way as to undermine its capacity to continue to perform its resource, waste-absorption and life-support functions on which the economies, societies and ways of life themselves depend. The contribution of the environment to human life may be expressed, as has been seen, in terms of *environmental functions* (Hueting 1980; de Groot 1992). A given environmental resource can serve many functions, but some uses preclude others. It is this competition between functions that gives the environment its essential economic characteristics.

Functions can be categorised into three groups, those that provide resources for economic activity; those that absorb wastes from economic activity; and those that provide services (e.g. climate, ozone-shielding, recreation, amenity) independently of or interdependently with human activity. Precisely analogously to economic sustainability, which requires that income is that quantity of consumption which can be sustained indefinitely (because the capital stock has been maintained), environmental sustainability requires that important environmental functions are used such that they can be sustained indefinitely (because the natural capital stock has been maintained). Perceptions of environmental unsustainability – the loss of important environmental functions – leading to economic and social unsustainability, is the reason that the idea of sustainability has become a major recent preoccupation in public policy.

Environmental unsustainability comes about when environmental functions which are important for human ways of life and welfare are not sustained or put at risk. Given the uncertainties involved in matters of sustainability, and the possibility of irreversibility, and of the incidence of very large costs, once environmental functions have been lost, the question of risk is crucial, not least because risk, through insurance premia for example, incurs present real, as well as hypothetical future, costs.

The simultaneous coincidence of uncertainty, irreversibility and possible large costs has long been recognised as an important consideration for environmental policy. Ciriacy-Wantrup's classic work (1952) prefigured many of the current concerns of sustainability with his development of the concept of 'the safe minimum standard'.

First Ciriacy-Wantrup (1952: 38ff.) identifies the existence of 'critical zones' for many, especially renewable, resources, where such a zone 'means a more or less clearly defined range of rates (of flow of the resource) below which a decrease in flow cannot be reversed *economically* under presently foreseeable conditions. Frequently such irreversibility is not only economic but also technological' (ibid.: 39) and, one may add with regard to extinguished species, biological. In the terminology being employed here, this means that the loss of environmental functions may be irreversible. The 'critical zone' concept is strikingly similar to that of the 'critical load' which is employed in modern environmental policy.

Then Ciriacy-Wantrup (ibid.: 88) identifies the possibility of 'immoderate losses' arising from environmental degradation, with respect to which: 'One important objective of conservation decisions is to avoid *immoderate* possible losses – although of small probability – by accepting the possibility of moderate ones – although the latter are more probable.' A decision rule which would achieve this is the 'minimax' criterion, which involves minimising maximum possible losses. The application of this criterion to resources characterised by critical zones leads Ciriacy-Wantrup (ibid.: chapter 18, 251ff.) to recommend the 'safe minimum standard' (SMS) as an objective of conservation (what today would be called environmental) policy: 'A safe minimum standard of conservation is achieved by avoiding the critical zone – that is, those physical conditions, brought about by human action, which would make it uneconomical to halt and reverse depletion'.

Despite the fact that the SMS approach was addressed in a practical way at those environmental problems, characterised by chronic uncertainty, and possible irreversibility and immoderate losses, for which more conventional economic approaches based on cost–benefit analysis are either inappropriate or infeasible, it made little impact on the literature of subsequent decades. The next substantial reference is Bishop (1978), by which time the Total Economic Value framework (mentioned earlier) of use, option and existence values had been developed, and Bishop relates this to the SMS. Bishop (1978: 17), concludes his article, which focused on endangered species, with the observation 'it is worthwhile to note that problems of irreversibility and uncertainty are not limited to endangered species, that the SMS approach may be applicable to a wider range of resource issues.'

Bishop (1993: 72) brings the SMS approach into the context of current environmental discourse by relating it to sustainability: 'To achieve sustainability policies should be considered that constrain the day to day operations of the economy in ways that enhance the natural resource endowments of future gener- ations, but with an eye towards the economic implications of specific steps to implement such policies.' Here the safe minimum standard has been converted into a sustainability standard. In the terms previously discussed, those activities that entail the possibility of irreversible effects and immoderate costs are now identified as environmentally unsustainable.

The SMS approach overcomes the problems of valuing environmental damages by proposing that policies that constrain or transform human activities towards environmental sustainability should not be considered in a normal benefit–cost framework but one which seeks to achieve the sustainability standard in a cost- effective way. Sustainability is accorded this pre-eminence as a policy objective because of the importance of environmental functions for human welfare, and because of the irreversibilities and large costs that may be associated with their loss. However, given that not all environmental functions can always be main- tained, what counts as an unsustainable loss of an environmental function, rather than a sustainable economic cost, is a matter of judgement which can only par- tially be resolved by science. Ethics and the attitude to risk also play a significant role here. It is important that the basis of judgement is articulated clearly, espe- cially as to who is responsible for the effects and who is bearing their costs, and differentiating the contributions played by science, ethics and risk acceptance or aversion.

In order to make the SMS approach operational, detailed sustainability stand- ards will need to be formulated for all the important environmental functions that are perceived to be at risk from human activities, but some general principles for these standards can be posited with regard to the generic functions of resource use, waste absorption and life support. Daly (1991: 44–45) has suggested four principles of sustainable development:

1   Limit the human scale (throughput) to that which is within the earth's carrying capacity.
2   Ensure that technological progress is efficiency-increasing rather than throughput-increasing.
3   For renewable resources harvesting rates should not exceed regeneration rates (sustained yield); waste emissions should not exceed the assimilative capacities of the receiving environment.
4   Non-renewable resources should be exploited no faster than the rate of creation of renewable substitutes.

These principles are among the rules that Turner (1993: 20–21) has formulated 'for the sustainable utilisation of the capital stock', the others of which are: correction of market and intervention failures; steering of technical change not only to increase resource-using efficiency but also to promote renewable substitutes for non-renewable resources; taking a precautionary approach to the uncertainties involved.

Of these rules, the correction of failures and the steering of technical change are more to do with achieving sustainability than defining standards for it; and in view of the complexity of applying the concept of carrying capacity to human activities, it seems desirable to express it more specifically in terms of those environmental problems that appear most pressing. Such considerations enable the Daly/Turner rules to be reformulated into a set of sustainability principles:

1   Destabilisation of global environmental features such as climate patterns or the ozone layer must be prevented. Most important in this category are the maintenance of biodiversity (see below), the prevention of climate change, by the stabilisation of the atmospheric concentration of greenhouse gases, and safeguarding the ozone layer by ceasing the emission of ozone-depleting substances.
2   Important ecosystems and ecological features must be absolutely protected to maintain biological diversity. Importance in this context comes from a recognition not only of the perhaps as yet unappreciated use value of individual species, but also of the fact that biodiversity underpins the productivity and resilience of ecosystems. Resilience, defined as 'the magnitude of the disturbance that can be absorbed before the system changes its structure by changing the variables and processes that control its behaviour' (Folke *et al.* 1994: 6) depends on the functional diversity of the system. This depends in turn, in complex ways, not just on the diversity of species but on their mix and population and the relations between the ecosystems that contain them. 'Biodiversity conservation, ecological sustainability and economic sustainability are inexorably linked; uncontrolled and irreversible biodiversity loss ruptures this link and puts the sustainability

of our basic economic-environmental systems at risk.' (Barbier *et al.* 1994: 41.)

3   The renewal of renewable resources must be fostered through the mainten-ance of soil fertility, hydrobiological cycles and necessary vegetative cover and the rigorous enforcement of sustainable harvesting. The latter implies basing harvesting rates on the most conservative estimates of stock levels, for such resources as fish; ensuring that replanting becomes an essential part of such activities as forestry; and using technologies for cultivation and harvest that do not degrade the relevant ecosystem, and deplete neither the soil nor genetic diversity.

4   Depletion of non-renewable resources should seek to balance the mainten-ance of a minimum life-expectancy of the resource with the development of substitutes for it. On reaching the minimum life-expectancy, its maintenance would mean that consumption of the resource would have to be matched by new discoveries of it. To help finance research for alternatives and the even-tual transition to renewable substitutes, all depletion of non-renewable resources should entail a contribution to a capital fund. Designing for resource-efficiency and durability can ensure that the practice of repair, reconditioning, re-use and recycling (the 'four R's') approach the limits of their environmental efficiency.

5   Emissions into air, soil and water must not exceed their critical load, that is the capability of the receiving media to disperse, absorb, neutralise and recycle them, nor may they lead to concentrations of toxins that cause unacceptable damage to human health. Synergies between pollutants can make critical loads very much more difficult to determine. Such uncertainties should result in a precautionary approach in the adoption of safe minimum standards.

6   Landscapes of special human or ecological significance, because of their rar-ity, aesthetic quality or cultural or spiritual associations, should be preserved.

7   Risks of life-damaging events from human activity must be kept at very low levels. Technologies which threaten long-lasting ecosystem damage should be foregone.

Of these seven sustainability principles, 3, 4 and, to some extent, 2 seek to sustain resource functions; 5 seeks to sustain waste-absorption functions; 1 and 2 seek to sustain life-supporting environmental services; 6 is concerned with other environmental services of special human value; and 7 acknowledges the great uncertainties associated with environmental change and the threshold effects and irreversibilities mentioned above. The principles are clearly an application of the SMS approach, but they seek to reinterpret it explicitly in the context of ensuring the sustainable use of important environmental functions.

Interpreting the SMS approach as the application of sustainability principles to derive sustainability standards to ensure the sustainable use of important environ-mental functions offers a practical methodology to address those environmental problems where uncertainty, irreversibility or possible large costs make the use of benefit–cost analysis problematic. The methodology was applied by the German Advisory Council on Global Change to recommend the necessary level of CO_2 reduction to the First Conference of the Parties on Climate Change in Berlin (WBGU 1995). Hueting *et al.* (1992, Appendix 2, 44–52) also used the methodology to indicate how sustainability standards can be derived for seventeen different environmental problems.

The mere identification of sustainability standards will not, of course, result in their adoption as the targets of public policy. Some such standards (for example, those required to preserve climate stability) would in any case require very substantial changes in social and economic activities and structures, and could only be achieved over a long time period if acute social and economic disruption were to be avoided. In some cases it might be that sustainability concerns would be politically over-ruled in favour of other economic or social priorities. Even in such cases, however, the advantage of having identified clear sustainability goals for environmental policy would be that the environmental trade-off against competing economic or social objectives would be clearly identified.

In fact, the sustainability principles are coming to be reflected in a number of international treaties, conventions and principles, including the Montreal Protocol to phase out ozone-depleting substances (1 above), the Convention on International Trade in Endangered Species and the establishment of World Biosphere Reserves to maintain biodiversity (2 above), the Second Sulphur Protocol to limit SO_2 emissions (5 above), and the Precautionary Principle, endorsed by the United Nations Conference on Environment and Development in Agenda 21, to limit environmental risk-taking (5 and 7 above). None of these international agreements was the outcome of detailed application of environmental evaluation techniques in a framework of cost–benefit analysis. They rest on a simple recognition that they represent the humane, moral and intelligent way for humans to proceed in order to maintain their conditions for life, and are argued for on that basis.

Estimating the Sustainability Gap

In order to understand environmental conditions, and to follow their evolution over time, it is necessary to choose environmental indicators, summary statistics which capture relevant information about the environmental condition in a way that is subject to unambiguous interpretation. Two common types of indicator are those which indicate the pressures on the environment from human activities (e.g. polluting emissions), and those which indicate the resulting state of the environment (e.g. pollutant concentrations).

Given an indicator of an environmental state or pressure that shows the current environmental position for some environmental issue, and a sustainability standard for the same indicator that shows a sustainable state, or the maximum pressure that is compatible with a sustainable state, with regard to that environmental issue, it is possible to determine a 'Sustainability Gap' ($SGAP_P$), in physical terms, between the current and a sustainable situation.

$SGAP_P$ indicates the degree of consumption of natural capital, either in the past or present, which is in excess of what is required for environmental sustainability. For the state indicators, the gap indicates the extent to which natural resource stocks are too low, or pollution stocks are too high. For pressure indicators, the gap indicates the extent to which the flows of energy and materials which contribute to environmental depletion and degradation are too high. $SGAP_P$ indicates in physical terms the extent to which economic activity is resulting in unsustainable impacts on important environmental functions.

$SGAP_P$ can give useful information as to the environmental impacts which need to be reduced, and by how much, but it does nothing to relate these impacts to the economic activities which are responsible for them, and which will need to be changed if the impacts are to be reduced. It would therefore seem desirable to link the sustainability indicators, and hence $SGAP_P$, to the national accounts, specifically by allocating the physical impacts responsible for the Sustainability Gap to the different sectors in the national accounts.

The sectoral disaggregation of environmental effects has been pioneered by the NAMEA (National Accounting Matrix including Environmental Accounts) system developed in the Dutch Central Bureau of Statistics. By 1996 this system covered the depletion of three types of natural resources – crude oil, natural gas and wood – and five types of environmental degradation: the greenhouse effect, depletion of the ozone layer, acidification, eutrophication, generation of wastes (Keuning 1996: 4–5). These environmental themes are related to seven aggregate economic sectors, which overall are responsible for generating the unsustainable environmental effects: agriculture, manufacturing (including oil refineries, chemical industries, basic metal industry, other manufacturing), electricity generation, construction, transport, services and other, and households. The NAMEA system has been recommended by the European Commission as the model for a European System of Integrated Economic and Environmental Indices (Keuning 1996: 2). Sweden and the UK have also decided to adopt NAMEA as the organising framework for their work.

The $SGAP_P$ indicators represent the physical improvements in the state of, or the reductions in the pressures exerted on, various environmental media and resources that are required in order to comply with the relevant sustainability standard. The NAMEA matrix can allocate the pressures among the various sectors that appear in the national accounts, but in order for the Sustainability Gap indicators to be compared directly with economic activity and output, they need to be given a monetary valuation. Consideration of technologies of abatement/ avoidance and restoration, by sector or with regard to different environmental media, enable, for each environmental theme, a series of cost curves to be built up, so that the cost of attaining each of the sustainability standards can be derived. These costs may then be aggregated to arrive at a full monetary figure for the Sustainability Gap ($SGAP_M$). Table 3.1 sets out the general schema, relating it to the normal input/output structure of the national accounts. It can be seen that in general abatement/avoidance costs are used to calculate the costs of reducing environmental pressures (e.g. emissions) to sustainable levels, while restoration

Table 3.1 Matrix for the construction of the Sustainability Gap

	Economic sectors	Environmental impacts		
		Current	Sustainable	SGAP
Economic sectors	Input/output tables	NAMEA pressure indicators		
Totals			physical	monetary[a]
Environmental quality		state indicators	physical	monetary[b]

Source: Ekins and Simon 1998: 161.

Notes
[a] Calculated using abatement/avoidance costs.
[b] Calculated using restoration costs.

costs are used to calculate the costs of restoring environmental states to sustainable levels. The details of how these calculations are performed are beyond the scope of this chapter. They are discussed in Ekins 1999.

$SGAP_M$ would be an expressive indicator of the potential of an economy, at a certain moment in time, to achieve environmental sustainability. It would reflect both the physical distances from environmental sustainability and the economic possibilities of reducing those distances. Over time, $SGAP_M$ would decrease if either the physical Sustainability Gaps decreased, or new technologies, processes or materials were developed which enabled those gaps to be reduced at lower cost in the future. $SGAP_M/GDP$, either in aggregate or for each environmental theme, would also be an interesting indicator with which to make inter-country comparisons of environmental efficiency, in much the same way as energy intensity (Energy Use/GDP) is currently used.

The search for sustainable development

Defining the goal of environmental policy as environmental sustainability has meant that environmental problems have come increasingly to be seen in terms of *unsustainability*: of technologies, industrial processes and, ultimately, of whole human ways of life, because they were undermining the resource and environmental base which sustained them. Towards the end of the 1980s the solution to unsustainability was posited as *sustainable development*.

The meaning of sustainable development has been subject to much debate, definition and redefinition (see Pearce 1989 *et al.*: 173–85, for a 'Gallery of Definitions' and Lélé 1991 for a critical review). This intellectual ferment has reinvigorated and added several new dimensions to the environmental economics discourse outlined in the Section 'Economics and the environment' in this chapter.

The first effect is an emphasis on physical laws and ecological realities as constraints within which economic systems must operate. Gone are the free-standing flows of exchange value, i.e. flows of money, which used to be the almost exclusive means by which economies were described. In their place are circular flows of materials, and linear flows of energy, that obey the laws of thermodynamics (1. Matter and energy are conserved; 2. Activity in a closed system increases its entropy, i.e. disorder). There is a new concern for ecologists' notions of eco-system resilience and stability, carrying capacity and the proportion of the earth's primary photosynthetic product that can sustainably be appropriated by humans (currently the figure is about 40 per cent of terrestrial primary product).

Second, the recognition of the importance of many environmental values which currently are only experienced as 'externalities', so that they are routinely and perhaps undesirably sacrificed for monetised economic values, has led to new efforts to try to take these values into account, some of which have been outlined above: through extended cost–benefit or multi-criteria analysis; by relating them to the national accounts; or by emphasising non-monetary use, option and exist-ence values and exploring new ways of either giving a money value to them or injecting them directly into the political decision-making process.

Third, the long timescales over which current environmental changes will make their effects felt has intensified concern for future generations and the ethics of reducing their environmental options. In particular this has been expressed in the debate over discounting. Normal discount rates in use (5 per cent–10 per cent) reduce effectively to zero any costs or benefits more than 30 years hence. Any discount rate above 1 per cent does the same beyond a 100 year time horizon. Yet this timescale is short compared with the likely effects of global warming, or the half-life of some radioactive materials. There are heated disagreements among environmental economists as to what extent the discounting of such effects is justified (Cline 1992 thoroughly reviews this issue).

Fourth, the meaning of 'development' has been subject to reappraisal. To what extent can it be approximated by increases in per capita income? If other aspects of human life should be taken into account in measuring 'development', such as perhaps health, education, environmental quality or human rights, what measures should be used for these other aspects? Should they be given a monetary value? If so, how, and if not, how should they be compared with monetary gains and losses? There has been a spate of efforts to address such questions, for example the United Nations Development Programme's annual Human Development Index (UNDP annual) and Daly and Cobb's (1989) Index of Sustainable Economic Welfare. Such works do not provide definitive answers as to the nature of development, but they can increase understanding of the concept.

Fifth, and related to the previous point, there is the question as to what extent environmental sustainability is compatible with indefinitely increasing economic growth, not only in theory but also in practice. It is indisputable that one of the driving forces of environmental destruction, and perhaps the ultimate cause in the past, has been the explosion of consumption in the industrial world following industrialisation. Four-fifths of the world's people have yet to experience this increase in consumption, but the evidence suggests that they wish to do so. It may be theoretically possible for them to expand their economies with a fraction of the damage caused by the early industrial economies at a comparable stage of development. It may be theoretically possible for old industrial economies to continue growing but, at the same time, to reduce their overall environmental damage. But in practice, and with few local exceptions, environments continue to deteriorate, sometimes disastrously, as with ongoing deforestation, land degradation, water depletion or the collapse of fisheries. Even in Europe, the European Commission opened its Fifth Action Programme on the Environment with the gloomy observation that 'many current forms of activity and development are not environmentally sustainable' (CEC 1992a: 4), as indicated by 'a slow but relentless deterioration of the environment of the Community, notwithstanding the measures taken over the last two decades' (CEC 1992b: 3). It is hard not to conclude that the term 'environmentally sustainable economic growth' continues to be a practical, if not a theoretical, contradiction in terms.

Sixth, although it is true that economic growth can be problematic environmentally, it is also true that poverty is no guarantee of sustainability. Of course, most of human existence has consisted of sustainable subsistence lifestyles with very 'poor' living standards compared with current industrialised lifestyles, so that generalisation about poverty always being bad for the environment is obviously wrong. In fact Broad (1994) has shown how poor people often act to protect the environment against commercial agents of its destruction. But the fact remains that today there are large numbers of very poor people extracting unsustainable livelihoods at the expense of forests, soils, or fisheries. It is hard to see how environmental sustainability can be achieved in poor countries without the generation of more secure livelihoods for these people whose poverty is currently contributing to environmental degradation. One hopeful possibility, expounded by Robert Chambers (1992) is that these more sustainable livelihoods could be the result of the regeneration of degraded land, thereby contributing to both economic and environmental goals.

Seventh, it is becoming increasingly perceived that moving economies towards environmental sustainability will depend on fundamental institutional change:

- establishing new, or reforming old, social and economic institutions;
- making new international agreements, such as those on climate change and biodiversity agreed at the Rio Summit;
- transforming existing institutions, such as the World Bank and International Monetary Fund, or the General Agreement on Tariffs and Trade (GATT) and its successor World Trade Organisation, to make them more sensitive to environmental concerns; and
- empowering communities and local authorities, acting at the level at which people actually experience environmental benefits or damages, to engage in sustainable resource management.

Finally, it has become recognised that environmental unsustainability cannot be tackled through environmental policy alone. Human uses of the environment are deeply rooted in economic activity and social customs. It is becoming increasingly obvious that very poor people and societies, desperate for economic development, are unlikely to use their environments sustainably. Nor is environmental sustainability likely to derive from economic decline or an absence of basic social cohesion and purpose. Such realisations lie behind the often-repeated statements that sustainability has economic and social, as well as environmental, dimensions. The achievement of environmental sustainability is likely to require economic and social progress as well, but such progress will have to be of a very different sort to the industrial development that has been largely responsible for present-day environmental problems. This new sort of progress, which has not yet been observed in practice on a significant scale anywhere in a post-industrialisation society, is what is termed 'sustainable development'.

In addition to providing a means of relating environmental, economic and social policy to each other, the idea of sustainability can also potentially address some of the environmental considerations noted in the Introduction which are beyond the scope of economic analysis. Most importantly, sustainability as a normative principle is clearly able to satisfy concerns about intergenerational equity: if important environmental functions are being sustained, then future generations' life chances deriving from such functions are not being reduced. Moreover, if desperate poverty is unlikely to lead to environmentally sustainable behaviour, then concern for environmental sustainability demands that such poverty be alleviated. This could help address some of the most pressing issues of intragenerational equity. Finally, the ability to measure progress towards environmental sustainability, along the lines outlined earlier, means that it can provide the guiding principle for environmental policy, which economics, as discussed earlier, is unable to do. Economic analysis could still play a very important role in terms of giving guidance as to how progress towards environmental sustainability could be achieved cost-effectively, i.e. at least cost, but the sustainability goal itself would not be the result of economic deliberation. In addition, of course economic analysis in its benefit–cost form will still have a useful role to play in that subset of environmental decision-making situations in which the environmental costs and benefits can be adequately expressed in money terms.

However, making sustainability principles operational will still have to take account of technological possibilities and the costs of complying with them, and these are likely to be reflected in the speed towards which sustainability standards are approached. In industrial societies the sectors that currently most conflict with sustainability principles are energy use, agriculture, some industrial processes, transport, waste-disposal and tourism. These are the areas to which the various theories of environmental economics and instruments of environmental policy – taxes, regulations, standards, tradable permits, treaties and the like – need to be applied if sustainability is to be approached. It is not yet clear whether such an approach would entail a new kind of industrial society or an alternative to it. Such uncertainty is unsettling. But it may be preferable to the kind of outcome, if present trends continue, envisaged by the World Resources Institute (WRI), when, in collaboration with both the Development and Environment

Programmes of the United Nations and on the basis of one of the worlds most extensive environmental databases, it concluded: 'The world is not now headed toward a sustainable future, but rather toward a variety of potential human and environmental disasters' (WRI 1992: 2).

References

Barbier, E.B., Burgess, J. and Folke, C. (1994) *Paradise Lost? The Ecological Economics of Biodiversity*, Earthscan, London.

Barnett, H. (1979) 'Scarcity and growth revisited' in Smith, V. (ed.) 1979 *Scarcity and Growth Reconsidered*, Johns Hopkins University Press, Baltimore.

Barnett, H. and Morse, C. (1963) *Scarcity and Economic Growth: The Economics of Natural Resource Availability*, Johns Hopkins University Press, Baltimore.

Bishop, R. (1978) 'Endangered species and uncertainty: the economics of a safe minimum standard', *American Journal of Agricultural Economics*, Vol. 60 (February), pp. 10–18.

Bishop, R. (1993) 'Economic efficiency, sustainability and biodiversity', *Ambio*, Vol. 22 No. 2–3 (May), pp. 69–73.

Broad, R. (1994) 'The poor and the environment: friends or foes?', *World Development*, Vol. 22, No. 6, pp. 811–22.

CEC (Commission of the European Communities) (1992a) Proposal for a Resolution of the Council of the European Communities, *Towards Sustainability: A European Community Programme of Policy and Action in Relation to the Environment and Sustainable Development*, Volume 1, Commission of the European Communities, Brussels.

CEC (Commission of the European Communities) (1992b) Executive Summary. *Towards Sustainability: A European Community Programme of Policy and Action in Relation to the Environment and Sustainable Development*, Volume 2, Commission of the European Communities, Brussels.

Chambers, R. (1992) 'Sustainable livelihoods: the poor's reconciliation of environment and development', In: P. Elkins and M. Max-Neef, (eds). *Real-Life Economics: Understanding Wealth Creation*. Routledge, London, pp. 214–29.

Ciriacy-Wantrup, S.V. (1952) *Resource Conservation: Economics and Policies*, University of California Press, Berkeley.

Cline, W. (1992) *The Economics of Global Warming*, Institute for International Economics, Washington DC.

Coase, R.H. (1960) 'The problem of social cost', *Journal of Law and Economics*, Vol. 3, pp. 1–44.

Cropper, M. and Oates, W. (1992) 'Environmental economics: a survey', *Journal of Economic Literature*, Vol. 30, June, pp. 673–740.

Daly, H.E. (1991) 'Elements of environmental macroeconomics' in Costanza, R. (ed.) 1991 *Ecological Economics: The Science and Management of Sustainability*, Columbia University Press, New York.

Daly, H. and Cobb, J. (1989) *For the Common Good: Redirecting the Economy Towards Community, the Environment and a Sustainable Future*, Beacon Press, Boston, and Green Print (Merlin Press 1990), London.

Dasgupta, P. and Heal, G. (1970) *Economic Theory and Exhaustible Resources*, Cambridge University Press, Cambridge.

de Groot, R. (1992) *Functions of Nature*, Wolters-Noordhoff, Groningen, Netherlands.

Diamond, P. A. and Hausman, J.A. (1994) 'Contingent Valuation: Is Some Number Better Than No Number?', *Journal of Economic Perspectives*, Vol. 8 No. 4 (Fall), pp. 45–64.

Ekins, P. and Simon, S. (1998) 'Determining the Sustainability Gap: national accounting for environmental sustainability' in Vaze, P. (ed.) 1998 *UK Environmental Accounts: Theory, Data and Application*, Office for National Statistics, London, pp. 147–67.

Ekins, P. (1999) *Economic Growth and Environmental Sustainability: The Prospects for Green Growth*, Routledge, London/New York.

Folke, C., Holling, C. S. and Perrings, C. (1994) 'Biodiversity, ecosystems and human welfare', Beijer Discussion Paper' Series No. 49, Beijer International Institute of Ecological Economics, Stockholm.

Hall, D. and Hall, J. (1984) 'Concepts and measures of national resource scarcity, with a summary of recent trends', *Journal of Environmental Economics and Management*, December, pp. 363–79.

Heal, G. and Barrow, M. (1980) 'The relationship between interest rates and metal price movements', *Review of Economic Studies*, Vol. 47, pp. 161–81.

Hotelling, H. (1931) 'The economics of exhaustible resources', *Journal of Political Economy*, Vol. 39, No. 2, pp. 137–75.

Hueting, R. (1980) *New Scarcity and Economic Growth*, North Holland, Amsterdam.

Hueting, R., Bosch, P. and de Boer, B. (1992) *Methodology for the Calculation of Sustainable National Income*, Statistical Essay M44, Netherlands Central Bureau of Statistics, Voorburg/Heerlen.

Jevons, W. (1965) *The Coal Question: An Inquiry Concerning the Progress of the Nation and the Probable Exhaustion of Our Coal Mines* (third edition, first edition published 1865), Augustus M. Kelley, New York.

Keuning, S. (1996) 'The NAMEA experience: an interim evaluation of the Netherlands' integrated accounts and indicators for the environment and the economy', paper presented to the International Symposium on Integrated Environmental and Economic Accounting in Theory and Practice, Tokyo, March 5–8, mimeo, Statistics Netherlands, National Accounts Department, Voorburg.

Lélé, S. (1991) 'Sustainable development: a critical review', *World Development*, Vol. 19, No. 6, pp. 607–21.

Lutz, E. (ed.) (1993) *Toward Improved Accounting for the Environment*, World Bank, Washington DC.

Malthus, T. (1970) *An Essay on the Principle of Population* (first edition, first published 1798), Penguin, Harmondsworth.

Malthus, T. (1974) *Principles of Political Economy: Considered with a View to their Practical Application* (second edition, first published 1836), Augustus M. Kelley, Clifton NJ.

Meadows, D.H., Meadows, D.L., Randers, J. and Behrens, W. (1972) *The Limits to Growth*, Universe Books, New York.

Meadows, D.H., Meadows, D.L., Randers, J. (1992) *Beyond the Limits*, Earthscan, London.

Mill, J.S. (1904) *Principles of Political Economy With Some of Their Applications to Social Philosophy* (sixth edition, first edition published 1848), Longmans, Green and Co., London.

Pearce, D. (1993) *Economic Values and the Natural World*. Earthscan, London.

Pearce, D., Markandya, A. and Barbier, E. (1989) *Blueprint for a Green Economy*, Earthscan, London.

Pearce, D. and Turner, R.K. (1990) *Economics of Natural Resources and the Environment*, Harvester Wheatsheaf, Hemel Hempstead, Herts.

Pigou, A.C. (1932) (4th edition) *The Economics of Welfare*, Macmillan, London.

Ricado, D. (1973) *The Principles of Political Economy and Taxation* (first edition 1817), J.M. Dent & Sons, London.

Smith, V. (1981) 'The empirical relevance of Hotelling's model for natural resources', *Resources and Energy*, Vol. 3, pp. 105–18.

Turner, R.K. (1993) 'Sustainability: principles and practice' in Turner, R.K. (ed.) 1993 *Sustainable Environmental Economics and Management: Principles and Practice*, Belhaven Press, New York/London, pp. 3–36.

UNDP (United Nations Development Programme), annual form (1990) *Human Development Report*, Oxford University Press, Oxford/New York.

UNSD (United Nations Statistical Division) (1993) *Integrated Environmental and Economic Accounting: Handbook of National Accounting*, Studies in Methods, Series F, No. 61, Interim Version, UN Statistical Division, New York.

WBGU (German Advisory Council on Global Change) (1995) 'Scenario for the derivation of global CO_2 reduction targets and implementation strategies', WBGU, Bremershaven.

WRI (World Resources Institute) (with UNDP and UNEP) (1992) *World Resources, 1992–93*, Oxford University Press, Oxford/New York.

4 Land use policy and sustainability

P. T. Kivell

Introduction

Land use and land use policy are absolutely central themes in the sustainability debate. This can be demonstrated in two broad ways. First, land is an essential ingredient of all human activity and a fundamental requirement for any form of production or development. There is almost universal evidence from recent history that population growth and the patterns of development and wealth creation that Western nations have pursued have created massively increased demands for space. Additionally the transport and telecommunications advances of recent years have resulted in a loosening of the constraints of distance, allowing both a dispersal and an increase in the scale of human land-using activities. These increased space demands, and the land use changes that have accompanied them, together with the risk of social polarisation between rich and poor over access to land, have had profound implications for environmental protection and management. Second, in Britain perhaps even more than elsewhere, the system and institutions of planning that promote and regulate both economic development and environmental protection have always been firmly based upon land use. Planning is centrally concerned with allocating land between competing uses and it is clear that when growing demands upon a fixed amount of land exist alongside increased environmental awareness and heightened concerns (however imprecisely formulated) about sustainability, it is a field where the potential for conflict is becoming more acute.

Currently the world is at an important threshold, with most estimates putting the proportion of people living in urban areas reaching 50 per cent by the millennium. In the developed Western world the proportion is already close to 75 per cent. In addition to this concentration of people, the world is also becoming more urban through the concentration of decision making and power in cities. For these reasons the present discussion will focus largely upon urban land, although, crucially, we must also recognise two important consequences of this global urbanisation of society. The first is the intensification of agriculture and the changes in crop-land world-wide which have been necessary in order to feed the growing urban populations, and the consequent impacts upon greenhouse gases and other matters that materially affect global sustainability. The second consideration is that the more people become concentrated into the built environment of cities, the more they become detached or isolated from the 'natural' physical environment and the less experience they have of the natural forces which shape and limit the land and its possible uses.

The nature of land

The issue is complicated by the peculiar nature of land itself. As Ratcliffe (1976), Holliday (1986) and others have pointed out, land as a factor of production is in an odd position compared with capital, labour and enterprise. Economic theory has traditionally had difficulty in knowing how to treat it, or has at least given it an equivocal role. As an economic good or resource it has some unusual characteristics, for example it is immobile and permanent, in general it is in fixed supply and nothing else can be substituted for it. Although agricultural fertility may be subject to laws of diminishing returns (Marshall 1920), within the span of human history land is not destroyed or used up, only changed and perhaps damaged. For over two hundred years there has been a view of land within many societies that there is a private right to treat it as an expendable market commodity. Now, prompted by the sustainability debate, there is a growing awareness of responsibility to future generations, although as Owens (1994) suggested there is little agreement on what constitutes the critical natural capital in land to be passed on to future generations.

It is also the case that the 'value' of land is not necessarily amenable to forms of economic analysis that may be appropriate for other commodities. Psychological and emotional values and notions of patriotism may deeply affect the way in which we view land and some have gone as far as to describe it as a 'living entity . . . its sanctity should be absolute' (Moss 1981). Land, and hence judgements about its sustainability, mean different things to different groups of people. To the lawyer a parcel of land represents a package of rights and obligations, to the economist and to the business sector it represents various forms of financial return or opportunity, to the owner of a small family farm it is a way of life, to the planner it is a basis for various forms of development and to the environmentalist it is both the physical stage upon which current environmental issues are played out and a heritage to be safeguarded for future needs. If this appears to be an unduly compartmentalised view of things, it is introduced in this way partly to underline the necessity for us to adopt new ways of looking at the role of land in modern society and partly to stress the need for fresh approaches to the social and economic expectations that we place upon it. Taking the argument rather further Stein and Harper (1996) have argued the need for a complete break with the moral basis of western civilisation and the search for a new, ethical environmental paradigm, which would include the way that we view land issues.

Sustainable land use

Although widely used, especially in a general sense, the notion of sustainability is still not easy to pin down, especially with regard to land use, and even more with regard to urban land. It is possible to find 'sustainable' development used as shorthand for efficient development, environmentally sympathetic development, small-scale development or even no development. Owens (1994) distinguished between a weak view of sustainability, which simply involves giving environmental considerations more weight in planning decisions, and a strong view, which recognises that environmental capacities place an eventual limit upon development. Most of the circumstantial evidence suggests that planners are happier to work with the weak view. Clearly sustainable development is an attractive concept, but giving it precise meaning in land use terms is not easy. Sustainability itself is capable of many definitions, Haughton and Hunter (1994), for example listed twenty-four alternatives. Some even argue that urban sustainability is a contradiction since cities necessarily consume the resources of a wider environment for their survival (Owens 1992). Up to a point this argument can be countered by the recognition that the city is a valuable resource itself and that, for a given population, the concentration of development within an urban cluster may minimise some environmental impacts.

In a practical sense the way in which we relate sustainability to the issue of land use planning will depend upon how we define the concept, and the alternatives from a planning viewpoint include the following:

- a condition that can be maintained indefinitely without progressive diminution in quality (Holdren *et al.* 1995);
- maintaining the capacity of natural ecosystems to support the human population over the long term (Alberti and Susskind 1996);
- the achievement of urban development aspirations, subject to the condition that the natural and man-made stock of resources are not so depleted that the long-term future is jeopardised (Breheny 1990);
- a city that is 'user-friendly' and resourceful in terms not only of its form and energy efficiency, but also its function, as a place for living (Elkin *et al.* 1991);
- sustainable development does not mean less economic development, or that every aspect of the present environment should be preserved unchanged, it means that decisions through society should be taken with proper regard to their environmental impact (Department of the Environment 1994).

Although sustainability and land use planning in cities is now clearly linked in principle, there is real confusion about the precise nature of the connection. There is a great deal of uncertainty about how sustainability can be implemented practically through the land use planning process (Tewdwr-Jones 1997) and there may even be a temptation to hook 'sustainability' onto any convenient noun (Merrett 1995). It is difficult to demonstrate that one form of urban development or land use pattern is generally superior to another because most land related issues are locally, or site, specific. It is also the case that after two centuries of urban/industrial development, planners in countries like Britain are not starting with a clean slate but have to adapt and incrementally change a huge stock of inherited physical development and social values. In some ways it is easier to demonstrate the precepts of sustainability in relation to limited issues such as urban parks, city farms or cycle-ways than it is to incorporate them into the mainstream core of land use planning and policy.

Urban issues

Concepts of sustainability can be applied to almost every aspect of urban development, from the broadest concerns of population growth or social equity to the most detailed aspects of the use of a particular parcel of land or the planting of trees in a city street. Many of these have been the subject of concerted programmes of research supported by the main funding councils (ESRC, EPSRC and NERC) since the early 1990s. In practical terms we can consider that a relatively limited number of issues dominate the current debate and these will be briefly reviewed below.

Carrying capacity

The idea that land can only support a given number of people, or a given level of human activity is well established. As urban populations and urban needs grow, so cities directly take more land. In England and Wales the urban area will have increased by approximately 105,000 hectares between 1981–2001 and in total 11 per cent of the land is now urbanised. In many West European countries approximately 2 per cent of agricultural land is lost to development each decade. The problem as far as cities are concerned is that the real environmental impact is often exerted indirectly at some distance from the city itself and is thus hidden from the user-group. With engineering and architectural advances very high population can be accommodated upon relatively small areas of land in urban areas. But these populations still need food, water and air and their waste products need to be disposed of. These functions can only be satisfied by using land outside the city, often at very great distances from it. The problem is exacerbated not simply by the need to accommodate more people, but by what Catton (1986) referred to as the need to accommodate 'larger' people, that is people who, especially in Western urban societies, are using per capita more and more energy and other resources.

The notion of a carrying capacity has been refined by Rees and Wackernagel (1996) into the calculation of an ecological footprint which is a measure of the land needed to support present populations and consumption levels. By this means they estimated that each Canadian, for example, needs 4.3 ha of land. This is almost three times their 'fair' earth-share. Vancouver needs a land areas 180 times its own administrative size to support its lifestyles and in the case of London the multiplier is 120, giving a total area for its support that exceeds the whole of

England and Wales. At a national level even the Netherlands, commonly cited as a country with well-developed environmental awareness and planning, has an ecological footprint 15 times the size of its national territory. Mostly these calculations focus upon the basic needs of food, material resources and CO_2 assimilation, so there is still scope for adding further land for amenity and aesthetic needs, or what Owens (1994) called post-material concerns.

Urban expansion

Although an important task for planners must remain the question of how best to deal with the existing stock of urban development, there is a huge challenge in meeting the land use needs of continuing urban expansion. No longer is this fuelled in the Western world mainly by simple population growth but by social, demographic and economic change. The ageing of populations and the decline of traditional, stable patterns of family structure have increased the number of separate households for a given population. New, low-density lifestyle preferences have been made possible by growing wealth and car-ownership. Many employment, retail and leisure activities have chosen to expand on attractive green-field sites rather than within existing urban areas. All of these forces have combined to create massive demands for building land and urban expansion around the fringes of cities and around nodes in the national motorway network.

Although the planning system (of which more later) does exert some controls upon this, it is far from effective in all instances and sustainability is only one of many criteria that it takes into account in approving the land use changes that accompany development. The last two decades, with a greater emphasis than previously upon demand-led development and free-market conditions, have witnessed far reaching land use changes on the urban fringe and elsewhere. To indicate the scale of this we can cite the government's own figures (Department of the Environment 1994) to show that in the mid 1990s annual planning approvals were given for:

- 150,000–200,000 new houses
- 2–2.5 million metre^2 of office floor-space
- 11.5–12.5 million metre^2 of industrial and warehouse floor-space
- 2–2.5 million metre^2 of retail floor-space.

In fact not all approvals are directly followed by development but these figures do give a crude indication of the high level of demand.

Equally crude figures have recently given added urgency to discussions of land use and sustainability as attempts have been made to predict the likely demand for new houses over the next few years. Government figures suggest that there will be a need for 4.4 million new houses between 1991–2016, a figure that will entail an urban expansion of some 172,000 ha, an area equivalent to 6½ times that of the present Birmingham or 16 times that of Bristol. Not surprisingly as this overall figure of 4.4 million has been translated in the detailed land and housing obligations placed upon local authorities (Table 4.1), doubts have been raised about the accuracy of the figures and the feasibility of this scale of development has been challenged. The potential impact has been viewed with greatest alarm in those counties which have hitherto been lightly urbanised, but where the current projections of growth are great. For example, in Cambridge, Somerset, Devon, Cornwall, Dorset and North Yorkshire the projections indicate an increase of approximately one-fifth in the amount of their urban land. Strong objections to the potential urbanisation of the countryside here and elsewhere have been raised by a range of national organisations, including the Council for the Protection of Rural England and *The Times* newspaper in its Greenfield Campaign, as well as by local communities and their MPs concerned about their own immediate environments. Partly as a result of these pressures, in February 1998, the Secretary for Environment, Transport and the Regions agreed that the old system of 'predict and provide' planning was no longer appropriate. Subsequently, forceful lobbying has achieved some changes and reductions in the original plans. In the South-East, for example, the South East Regional Planning Conference announced in late October 1998 a revised target for 862,000 new houses, compared with a figure of 1.1 million which the government had forecast for the region.

Urban form

One of the most persistent and powerful of forces shaping urban development in recent years has been that of decentralisation, leading, especially in North America, to low-density, sprawling cities. The apparently innocent, individual choices that people have made in opting for low-density, suburban lifestyles have had damaging cumulative effects. The process has been criticised for causing imbalances in the job and housing markets, a profligate use of land and increased land costs, traffic congestion and additional journey needs, a loss of community spirit, a loss of town centre vitality, excessive energy use, air and water pollution and other damage to the environment. As Hall (1997) suggested, this has resulted in cities moving away from rather than towards sustainable patterns, especially in North America, Britain and the rest of Europe and there is no doubting that counter-urbanisation has been, and remains, a very powerful trend.

Researchers including Banister (1992) and Newman and Kenworthy (1989) have drawn attention to the greater energy costs for transport in low density cities where there is a reliance on private motor transport, and it is clear that traffic congestion adds significantly to pollution problems in many cities, including notably London, Paris and Mexico City. Population densities in typical US metropolitan areas are low (13–14 people/ha), in European cities they are commonly 3 or 4 times as high and in Hong Kong, Singapore and Tokyo they are around 12 times US levels. Although there have now been several studies of the connection between urban densities, land use and travel generation, the overall evidence remains ambiguous. (Royal Commission on Environmental Pollution 1994.)

In consequence there has been much recent debate about what forms of urban

Table 4.1 Projections of urban growth, by county (England), 1991–2016

County	Urban growth[a] (hectares)	Rate of change to urban uses[b] (%)	Rate of urban growth[c] (%)	Level of urbanisation 2016[d] (%)
Avon	3,200	2.3	12.8	20.5
Bedfordshire	2,700	2.2	16.6	15.3
Berkshire	3,000	2.4	13.0	20.7
Buckinghamshire	4,300	2.2	17.2	15.3
Cambridge	5,900	1.7	21.3	9.8
Cheshire	4,500	1.9	14.2	15.4
Cleveland	1,000	1.7	6.6	26.8
Cornwall	4,200	1.2	19.4	7.1
Cumbria	2,400	0.4	12.0	3.3
Derbyshire	3,900	1.5	12.9	12.7
Devon	7,700	1.1	20.7	6.7
Dorset	4,400	1.6	19.2	10.1
Durham	2,400	1.0	14.6	7.6
East Sussex	2,700	1.5	13.8	12.2
Essex	7,000	1.9	15.1	14.4
Gloucestershire	2,700	1.0	13.3	8.7
Hampshire	6,600	1.7	13.1	15.0
Hereford & Worcester	4,400	1.1	17.1	7.5
Hertfordshire	2,800	1.7	9.0	20.1
Humberside	3,700	1.1	12.7	9.3
Isle of Wight	100	0.1	0.5	12.3
Kent	6,000	1.6	12.6	14.4
Lancashire	5,100	1.6	14.9	12.6
Leicestershire	5,200	2.0	18.9	12.8
Lincolnshire	4,300	0.7	13.7	6.0
Norfolk	4,500	0.8	12.4	7.6
North Yorkshire	5,300	0.6	19.1	3.9
Northamptonshire	3,800	1.6	18.7	10.2
Northumberland	1,900	0.4	15.6	2.7
Nottinghamshire	4,100	1.9	13.2	15.8
Oxfordshire	4,200	1.6	18.4	10.2
Shropshire	3,200	0.9	17.5	6.1
Somerset	4,400	1.3	20.8	7.3
Staffordshire	3,600	1.3	11.5	12.7
Suffolk	4,600	1.2	16.6	8.5
Surrey	2,200	1.3	5.9	23.5
Warwickshire	3,200	1.6	17.2	10.8
West Sussex	2,700	1.3	10.8	13.7
Wiltshire	4,500	1.3	18.6	8.1
Greater London	5,800	3.6	5.0	77.1
Greater Manchester	2,500	1.9	4.5	44.6
Merseyside	900	1.3	2.9	48.3
South Yorkshire	2,000	1.3	5.9	22.8
Tyne & Wear	1,700	3.0	6.9	46.2
West Midlands	1,600	1.7	2.6	66.0
West Yorkshire	4,700	2.3	9.9	25.2

Source: DoE, http://www.environment.detr.gov.uk/epsim/ems0804.htm (22 January 1999).

Notes
[a] The area projected to change net from rural uses to urban uses.
[b] The percentage of the total area projected to change net from rural uses to urban uses.
[c] The area projected to change net from rural uses to urban uses, expressed as a percentage of the area of land in urban uses in 1991.
[d] The percentage of the total area which is projected to be in urban uses in 2016.

development and land use might be more sustainable. Much of this argument has been conducted around the idea of building more compact cities, with higher densities and more mixed patterns of land use where reduced travel needs and a greater use of public transport can lead to major energy savings. Sprawling urban fringes need to be controlled and renewed emphasis needs to be placed upon redevelopment and in-filling of gaps in the existing urban form. In other words the urban development and planning paradigm needs to move away from much of what has been the common practise in British, and more especially in American cities, over the past generation.

Land use planning in post-war British and American cities often followed the orthodoxy of separating land uses into distinctive zones dominated by residential, industrial or commercial uses. Although there are some desirable characteristics of this pattern, it also generates the need for many relatively long distance urban journeys, for example to work or shop, and leads to a mono-functional sterility of land use. The case for small-scale, mixed land uses was promulgated by Jane Jacobs in the 1960s, and largely ignored by planners, but it has recently been revived. Coupland (1997) for example, suggested that it will lead to more attractive, diverse cities, that are safer to live and work in and where residents will be able to lead more sustainable lifestyles. Similar arguments have been put forward by Nystrom (1995) who linked the need for more mixed land uses with the reduction of transport use and the adoption of ecological principles in planning for sustainable cities. There are, of course contrary arguments, particularly in terms of possible conflicts between incompatible land uses, and the property development industry has often been wary of land uses that are too mixed.

A range of alternative urban forms confront the planners and policy makers. Haughton and Hunter (1994) summarised these in terms of four main groups which are listed below, with additional comment.

1   *Balanced integration with the regional natural environment.* This could include a return to Ebenezer Howard's Garden City concept of almost a century ago, and perhaps it illustrates how far-sighted he was. Settlements would be of modest size, nature is accommodated through green belts and green corridors and social and community needs are high on the agenda.

2   *Urban concentration and the control of nature.* Here high density cities are based upon new building materials and technologies in forms often associated (sometimes erroneously) with the work of Le Corbusier. Unfortunately the experience of high density, compact urban developments has not been good in many instances. Both Breheny (1992) and Owens (1986) have raised doubts about the efficacy and the desirability of (extreme) compactness and there is a clear risk that disproportionately high density settlements could prove unattractive and create fresh pressures for outward movement of people and activities from the cities.

3   *Concentrated decentralisation* whereby a 'galaxy' of separate, medium-sized settlements, separated by green space, make up a regional city. This concept is along the lines suggested some time ago by Gruen (1964) and Lynch

(1981), although the idea of concentrating development at nodes along a public transport corridor has more recently been espoused by both the Commission of the European Communities (1996) and Hall (1997).

4   *Decentralisation, dispersal and assimilation within the natural environment.* This represents a very scattered, low-density form of development in which distinctions between urban and rural all but disappear. In the high technology variant envisaged by Wright (1974) extended utility provisions and long-distance travel patterns would have made the arrangement environmentally inefficient, but other variants have proposed small-scale, low-technology, self-sufficient settlements with a much higher sustainability rating.

As already suggested the evidence on the connection between sustainable development and different urban forms is mixed, but in a wide ranging review Breheny (1992) found a glimmer of consensus around the following:

- Urban containment policies should continue to be adopted and the decentralisation process slowed down.
- Extreme compact city proposals are unrealistic and undesirable.
- Various forms of 'decentralised concentration', based around single cities or groups of towns may be most appropriate.
- Inner cities must be rejuvenated, thus reducing further losses of people and jobs.
- Urban (or regional) greening must be promoted.
- Public transport must be improved both between and within all towns.
- People-intensive activities must be developed around public transport nodes.
- Mixed uses must be encouraged in cities and zoning discouraged.
- Combined heat and power schemes must be promoted.

Transport

This is not the place for a lengthy discussion of urban transport, and in any case it has been referred to above, but in so far as it has important impacts upon urban sustainability it needs consideration. The city originally assumed a compact form largely because transport was slow, expensive and inefficient. With the rise of modern transport technologies, especially the motor vehicle, the city has been enabled to spread outwards and in the post-war years almost all urban functions, including housing, employment and retailing have decentralised. We have long past the point when it is possible to say confidently which is cause and which is effect, for as the motor car facilitated the dispersal of the city, it became necessary in turn for people to resort more and more to private transport in their day-to-day lives, and for there to be a greater use of motor transport to service and provide goods for the city. The planning of transport and land use is disjointed and the resulting land use pattern has progressively disadvantaged the operation of efficient public transport systems. Not only has the car been the prime agent of urban dispersal, but it also has direct impacts upon land use, taking up approximately one-third of the land in the typical American city.

Derelict and damaged land

Within the rapidly changing pattern of human activity in cities, land often becomes damaged, despoiled or wasted in various ways. At the simplest level of definition it is not sustainable for humans to damage land and then abandon it before moving on to develop elsewhere. As Moss (1981: 2) suggested 'we should not waste a single acre. But we do'. He went on to accuse society of profligacy in taking land for new, low density development whilst leaving behind extensive dereliction. It was clear from the late 1970s onwards that vacant and derelict land was a serious problem in many cities (Kivell 1993, 1999). In 1993 a Department of the Environment survey revealed a total of 39,600 ha of dereliction in England and, from 1990 figures, Scotland was shown to possess a further 8,297 ha. Apart from the obvious misuse of a resource that derelict land represents, its presence in concentrated local pockets has contributed to making life in the city unattractive.

Derelict land can be reclaimed for further beneficial use and substantial programmes of land restoration have taken place in British and other European cities. It is now clearly understood that the recycling of land can play a valuable role in protecting green-field sites from development, and in improving the urban environment. In recent years almost half of all new housing in British cities has been built on recycled land and there is now a high level of public interest and a clearly expressed government policy to use brown-field sites where possible.

Planning and policy responses

There has long been a strong protectionist theme running through British attitudes to land and the notions of stewardship of land resources and responsibility to future generations have traditionally been central to the land use planning process. This thread remains in place today, although as we shall see below the main reason for it has changed profoundly.

The new planning system of 1947 (which still forms the backbone of planning today) was largely predicated upon the basis of controlling the wasteful and uncaring use of land. This was shaped by events of the time, including urban squalor, dereliction and sprawl in the pre-war years, coupled with the needs to reclaim the agricultural wastes of the inter-war depression and the need to secure our own food supplies, the fragility of which had been exposed during the 1939–45 war. Land saving and urban containment became major tenets of the planning system from 1947 until the early 1980s, largely with a view to preserving the 'countryside' and, more especially, our agricultural land. Although major achievements can be recorded, not all commentators agreed that planning was doing an effective job in respect of land use (Coleman 1977, 1986).

A number of important publications on land use and land policy appeared in the 1970s and early 1980s (for example, Denman and Prodano 1972; Ratcliffe

1976; Lichfield and Darin-Drabkin 1980; Best 1981), and although none of these dealt with sustainability *per se* (after all the word was not current then), they all implicitly covered many of the key issues of sustainability. Best (1981) in particular was concerned with how many people and activities the land of Britain could support.

By the 1980s the strong protectionist thread continued to run through attitudes to land use planning, but by now the reasons for this were changing. In particular the availability of cheap food from international sources, the taste of the British consumer for more exotic foodstuffs and the apparent food surpluses with the European Community reduced some of the pressure to protect farmland in Britain. Instead strategies for managing land use and environmental change became more concerned with social welfare and environmental quality. The protectionist theme was now based much more upon environmental, aesthetic and amenity functions and it became bolstered by an alliance of old style conservatives and new style environmentalists. At the same time a growing middle class was attracted by the actuality, or at least the future dream, of a widespread 'countryside' lifestyle. From this complex set of processes sustainability emerged as an important issue.

The planning system was under enormous pressures at this time, and once again it did not always cope well. The pro-development era of the 1980s was largely run by an over-centralised planning system that Healey *et al.* (1988) claimed was politically unsustainable, although they also recognised that to swing too far the other way, towards highly localised planning, might be equally ineffective since individual districts do not necessarily act in ways that safeguard their own (or the country's) long-term interests. Cullingworth (1997) too was critical of land use planning and castigated it for its inability to deal with interrelated problems (such as housing, transport and town centres) and the fact that it had become too negative, too protectionist.

The 1980s was clearly a turbulent decade for planning. It was also the decade when the notion of sustainability was slowly being incorporated into British land use planning, both in principle, and, much more uncertainly, in practise, although it was not until the 1990s that the real discussions began. In 1990 the publication of the European Commission's *Green Paper on the Urban Environment* (Commission of the European Communities 1990) and the UK government's *This Common Inheritance* (Department of the Environment 1990) both acknowledged the role of land use planning in helping to achieve sustainable development. This was seen particularly through the connection between transport, land use, urban form and growth and the desirability of planning for mixed urban functions and land uses in place of strict zoning.

Four years later these links, and some of the strategies to be associated with them were spelled out more clearly in a White Paper entitled *Sustainable Development: the UK Strategy* (Department of the Environment 1994). This document stressed that the UK land use planning system had always existed to weigh the interests of economic development and environmental protection and went on to summarise three important principles of sustainability:

- To balance competing demands for a finite quantity of land available and to optimise development of vacant urban land and to reclaim derelict land;
- To protect the countryside for agriculture, landscape, wildlife and recreation; and
- To maximise access to facilities for individuals and businesses whilst minimising the amount of travel required.

In suggesting that urban growth should be encouraged as the most sustainable settlement form, the document acknowledged that compact cities use less land, enable lower energy consumption, reduce the need for travel and also help to enhance the quality of urban community life. Green belts were identified as important mechanisms for the delivery of compact cities and the reclamation of derelict land was also to be a priority in the efficient use of land. Undoubtedly the unwillingness to be any more prescriptive about the precise form that new settlement patterns might take was influenced by the Department of the Environment's (1993) own commissioned report on alternatives which had not been able clearly to demonstrate the superiority of one form over another, largely because the real problems are not so much general ones as local and site specific.

Whilst identifying many of the issues, and giving the planning system the key role in delivering land use compatible with the aims of sustainable development, and encouraging local authorities to incorporate environmental considerations more fully into their development plans, the White Paper was disappointingly lacking in practical advice.

In practice, some of this advice was to come through revised Planning Policy Guidance Notes (PPGs) issued by the Department of the Environment, a device that the government seemed to be using increasingly in order to give a steer to the direction of land use planning. The most relevant of these PPGs were as follows:

- PPG 1 (1988, revised 1992 and 1996) *General Policy and Principles* which established sustainability as a key requirement of the planning system;
- PPG 3 (1988, revised 1992) *Housing,* which included sustainability as a feature of new housing development;
- PPG 6 (1988, revised 1993 and 1996) *Town Centres and Retail Developments,* which addressed the vitality of town centres and the threat posed by out of town shopping;
- PPG 12 (1988, revised 1994 and 1996) *Development Plans and Regional Planning Guidance,* which reinforced the sustainability message in the preparation of development plans; and
- PPG 13 (1988, revised 1994) *Transport,* which advocated better integration of transport and land use planning.

Although these guidance notes largely concerned the direction of relatively localised planning, it was clear that the UK government was responding to an international agenda for sustainability. There was for example clear complementarity between some (but not by any means all) of the evolving UK strategy and the themes argued by the European Community Expert Group on the Urban Environment culminating in their report on *European Sustainable Cities* (Commission of the European Communities 1996). Although this document was as much about institutional structures as about practical planning, it undoubtedly provided an important stimulus for sustainability. In part it emphasised the importance of the local scale, by endorsing local Agenda 21 processes and by recognising the diversity of local problems and solutions, but it also addressed some of the really big planning issues. For example in chapter 7 it encouraged ecologically based approaches and a move away from a narrow land use focus, it suggested that planning should be 'supply-limited' rather than 'demand-driven' and it advocated that planners should not always seek to balance the benefits of development against costs to the environment but should instead define environmental capacities and prevent them from being breached.

Another international forum, through which much of the UK progress to date was summarised, was represented by the report to Habitat II (Department of the Environment 1996). Here the incorporation of sustainability into mainstream planning was endorsed in the following terms.

> The Town and Country Planning system is the framework within which the development and use of land is determined. It provides a structure within which economic, social and environmental considerations can be weighed to help secure sustainable development.
>
> (p. v)
>
> The government aims to ensure that the sum total of decisions in the planning field, as elsewhere, should not deny future generations the best of today's environment. The principles of sustainable development underlie Government planning guidance. . . .
>
> (p. vi)
>
> . . . the government will continue to ensure that the principles of sustainable development are taken fully into account in the operation of the land use planning system.
>
> (p. vi)

It went on to outline four crucially important planning objectives for sustainability:

1   to contain urban development;
2   to meet the need for new housing development in a sustainable way;
3   to promote viable town centres; and
4   to promote integrated land use and transport planning.

Much of the rest of the report was devoted to examples of sustainable development in the field of settlement patterns, transport, local economic development, environmental protection and health and urban quality. The conclusion was that 'The Government believes that land use planning has served the country well' (p. v), a sentiment with which not everybody would agree.

Conclusion

By 1996 it was clear that the UK government was committed to the incorporation of sustainability into mainstream planning. At a general level some messages were fairly clear and consistent, for example mixed land uses were felt to be beneficial, traffic generation and transport modes needed to be controlled and out of town shopping centres were likely to be regarded less favourably. Other messages however were either unclear or shifting, for example the policy on estimating and accommodating future housing needs, and everywhere there was a lack of clarity about precisely how sustainable land uses were to be implemented.

Almost certainly 1998 will be judged an important year for bringing sustainability more fully into land use planning and development. A number of important programmes were put into place. For example in September 1998 the Department of Environment, Transport and the Regions (DETR) announced the CLAIRE initiative (Contaminated Land – Applications in Real Environments) and also produced a Good Practice Guide for Sustainable Regeneration (DETR 1998a). Unfortunately a major statement on Modernising Planning (DETR 1998b) once again used 'sustainable' as little more than a shibboleth, but it did promise the announcement of new policies on two of the most crucial issues for sustainable planning, i.e. housing and transport, over the next year. These will be eagerly awaited as a real test of the government's commitment to sustainable land use planning.

In 1998 the government also published a guide to good practice in planning for sustainable development which went some way towards providing the practical advice that had previously been absent (DETR 1998c). This touches upon many issues, but essentially it gives advice on how to focus development within existing urban areas, on increasing housing densities whilst enhancing quality, on increasing the amount of development on brown-field sites and it proposes the development of urban villages and other types of mixed land uses. On transport, it suggests that authorities should concentrate settlement development within corridors and it puts forward suggestions for reducing car dependence.

But of course sustainable land use planning is not solely a matter for governments. One of the abiding problems of planning and development is that whilst most people want the benefits brought by industrial, commercial, housing and leisure developments, they don't want them in their immediate locality. The current resistance by many communities to new housing and new settlements, and the government's electoral sensitivity to this illustrates the nature of the problem. Government undoubtedly has a responsibility to take the lead with its policies and planning for sustainability, but there is still a long way to go in shaping public opinion, a process that has been started with the consultation document issued in 1998 (DETR 1998d).

The important tasks for the next decade include the business of shaping public opinion more widely, not just amongst a narrow group of people with '(G)green' interests but amongst groups with particular interests in the building and development industries, and of course, amongst the general public. At the same time the organisational framework must be put right, especially through the integration of land use planning and environmental planning, not merely as separate parts of a planning system but as the unified core of the system itself. Perhaps the ultimate challenge comes from recognising that difficult though it may be in the short term for us to adopt genuinely sustainable urban land use planning, in the long term the problems of rejecting it would be even more difficult to handle.

References

Alberti, M. and Susskind, L. (1996) 'Managing urban sustainability: an introduction to the special issue', *Environmental Impact Assessment Review*, 16/4: 213–22.

Banister, D. (1992) 'Energy use, transport and settlement patterns', in M. Breheny, (ed.) *Sustainable Development and Urban Form*, 160–81, London, Pion.

Best, R. (1981) *Land Use and Living Space*, London, Methuen.

Breheny, M. (1990) *Strategic Planning and Urban Sustainability*, Proceedings of the 1990 Town and Country Planning Association Annual Conference, London, TCPA.

Breheny, M. (1992) (ed.) *Sustainable Development and Urban Form*, London, Pion.

Catton, W. (1986) 'Carrying capacity and the limits to freedom', *Paper for Social Ecology Session 1, XI World Congress of Sociology*, New Delhi.

Coleman, A. (1977) 'Land-use planning: success or failure?' *Architect's Journal*, 165, 91–134.

Coleman, A. (1986) 'Research maps as an influence on public policy', in P.T. Kivell and J.T. Coppock, (eds) *Geography, Planning and Policy Making*, Norwich, Geobooks.

Commission of the European Communities (1990) *Green Paper on the Urban Environment*, EUR 12902, Brussels.

Commission of the European Communities (1996) *European Sustainable Cities, Final Report*, DGXI, Brussels.

Coupland, A. (1997) (ed.) *Reclaiming the City: Mixed Use Development*, London, E. & F.N. Spon.

Cullingworth, J.B. (1997) 'British and use planning; a failure to cope with change', *Urban Studies*, 34/5–6: 945–60.

Denman, D.R. and Prodano, S. (1972) *Land Use*, London, Geo Allen & Unwin.

Department of the Environment (1990) *This Common Inheritance: Britain's Environmental Strategy*, Cmnd. 1200, London, HMSO.

Department of the Environment (1993) *Alternative Development Patterns: New Settlements* (edited by M. Breheny, T. Gent and D. Lock), London, HMSO.

Department of the Environment (1994) *Sustainable Development: The UK Strategy*, Cmnd. 2426, London, HMSO.

Department of the Environment (1996) *Sustainable Settlements and Shelter, the United Kingdom National Report*, Habitat II, London, HMSO.

Department of the Environment, Transport and the Regions (1998a) *Sustainable Regeneration: Good Practice Guide*, http://www.regeneration.detr.gov.uk/ sustainable/guide/index.htm (5 October 1998).

Department of the Environment, Transport and the Regions (1998b) *Modernising Planning*, http://www.planning.detr.gov.uk/modern/1.htm (24 February 1998).

Department of the Environment, Transport and the Regions (1998c) *Planning for Sustainable Development: Towards Better Practice*, London, HMSO.

Department of the Environment, Transport and Regions (1998d) *Sustainable Development: Opportunities for Change: Consultation Paper on a Revised UK Strategy* http://www.environment.detr.gov.uk/sustainable/consult1/index.htm (4 February 1998).

Elkin, T., McLaren, D. and Hillman, M (1991) *Reviving the City: Towards Sustainable Urban Development*, London, Friends of the Earth.

Gruen, V. (1964) *The Heart of our Cities*, New York, Simon and Schuster.

Hall, P. (1997) 'The future of the metropolis and its form', *Regional Studies*, 31/3, 211–20.

Haughton, G. and Hunter, C. (1994) *Sustainable Cities*, London, Regional Studies Association.

Healey, P., McNamara, P., Elson, M. and Dock, A. (1988) *Land Use Planning and the Mediation of Urban Change*, Cambridge, Cambridge University Press.

Holdren, J.P., Daily, G.C. and Ehrlich, P.R. (1995) 'The meaning of sustainability', in *Defining and Measuring Sustainability, the Biophysical Foundations*, M. Munasinghe and S. Walter (eds), Washington, World Bank.

Holliday, J.C. (1986) *Land at the Centre*, London, Shepheard-Walwyn.

Jacobs, J. (1962) *The Death and Life of Great American Cities*, London, Jonathan Cape.

Kivell, P.T. (1993) *Land and the City*, London, Routledge.

Kivell, P.T. (1999) (in press) 'Derelict and vacant land', in M. Pacione (ed.) *Applied Geography*, London, Routledge.

Lichfield, N. and Darin-Drabkin, H. (1980) *Land Policy in Planning*, London, Geo Allen & Unwin.

Lynch, K. (1981) *Good City Form*, Cambridge Massachusetts, MIT Press.

Marshall, A. (1920) *Principles of Economics*, London, Macmillan.

Merrett, S. (1995) 'Planning in the age of sustainability', *Scandinavian Housing and Planning Research*, 12/1:5–16.

Moss, G. (1981) *Britain's wasting acres*, London, Architectural Press.

Newman, P.W.G. and Kenworthy, J.R. (1989) 'Gasoline consumption and cities – a comparison of US cities with a global survey', *Journal of the American Planning Association*, 55: 24–37.

Nystrom, L. (1995) 'The diversity of the urban environment', *Scandinavian Housing and Planning Research*, 12/4: 223–29.

Owens, S. (1986) *Energy Planning and Urban Form*, London, Pion.

Owens, S. (1992) 'Energy, environmental sustainability and urban form', in M. Breheny (ed.) *Sustainable Development and Urban Form*, 79–105, London, Pion.

Owens, S. (1994) 'Land, limits and sustainability: a conceptual framework and some dilemmas for the planning system', *Transactions of the Institute of British Geographers*, 19/4: 439–56.

Ratcliffe, J. (1976) *Land Policy*, London, Hutchinson.

Rees, W.E. and Wackernagel, M. (1996) 'Urban ecological footprints; why cities cannot be sustainable and why they are the key to sustainability', *Environmental Impact Assessment Review*, 16/4–6:223–48.

Royal Commission on Environmental Pollution (1994) *Transport and the Environment*, Oxford, Oxford University Press.

Stein, S. and Harper, T.L. (1996) 'Planning theory for environmentally sustainable planning', *Geography Research Forum*, 16: 80–101.

Tewdwr-Jones, M. (1997) 'Plans, policies and inter-governmental relations: assessing the role of national planning guidance in England and Wales', *Urban Studies*, 34/1: 141–62.

Wright, F.L. (1974) 'City of the future', in A. Blowers, C. Hamnett and P. Sarre (eds) *The Future of Cities*, London, Hutchinson.

5 Environmental health and sustainability

K. T. Mason

Introduction

The discussion of sustainability has rarely included serious deliberation about the relationship between health and survival and, more particularly, between health and the individual's life chances. In this chapter I examine environmental health as a critical component of thinking about sustainability, by placing it at the fore-front of a concern with life chances and, more indirectly, with the quality of life.

It is generally accepted now that the incidence and severity of disease and ill-health, and more widely, the quality of life, is influenced as much by where you live as it is by how you live, and that differentials in health and well-being can be attributed, in part a least, to contextual environmental circumstances. Simply put, the variable quality of the environment – be it the physical environment repre-sented by air or water quality or the condition of housing, or the social environ-ment represented by levels of social exclusion or the fear of crime – impinges upon the physical and psychological condition of the resident population. Central to this thinking is the notion that the exposure of humans to environmental hazards is likely to result in an adverse health outcome.

In this context 'health' needs to be viewed as being not simply an absence of disease but rather a state of complete physical, social and mental well being, with strong associations with the notion of 'quality of life'. The point here is that many of the factors which combine to shape the quality of life are health- and environment-related – noise, air quality, housing standards, occupational health, food safety, etc. Although many environmental health hazards may be entirely natural in origin there are many more that derive from human activities – indus-trial or agricultural pollution for instance. In such cases the threat to health only occurs because of the existence of some motivating force which creates the hazard in the first place. Such motivation is generally found in economic development – a process reliant upon accelerated economic activity and one that is commonly accompanied by a wide range of occupational or other pressures upon the environment. In health terms, these pressures may range from the release of increased levels of industrial pollutants and subsequent heightened contamin-ation of the physical environment, through to expanded household consump-tion and the problems of waste disposal, to various impacts upon the social environment – such as the increased rates of unemployment that sometimes come with technological change.

Although it is primarily these negative impacts upon health that will be the subject matter of this chapter we should not lose sight of the fact that the prosperity and sense of achievement that comes with economic growth brings with it massive benefits to health which arguably outweigh the problems arising from the accompanying health-related hazards. Here, the concept of scale becomes important because many of the environmental health hazards arising from economic activity operate on a local scale rather than on the wider population that derives benefit from the health improvements that economic growth facilitates. We should also remember that the interaction between health and development is actually a two-way process. This is the case because the benefits to health that can accrue from economic progress are themselves likely to facilitate further economic development, for the simple reason that healthy people are more productive economically. So, not only does economic development impact upon health, but health issues themselves will have a major bearing upon the speed of economic progress – investing in health can be a powerful tool for accelerating development (World Bank 1993). From the above then, we can recognise three important relationships which together conspire to define or change the health of a population – health and environment, environment and economy, and economy and health (Labont 1991). By reason of this three-way interaction progress in any one of these individual domains will have implications for the other two, and the creation of policy supportive of health therefore requires an integrated strategy encompassing the different policy domains. This is perhaps where the notion of sustainability is best employed – as an integrative concept which brings together the interests of all sectors of society and facilitates broad action incorporating health, economic and environmental concerns.

This chapter will consider sustainability issues in a health context and will explore the links between health and environment and the role of economic activity in shaping individual life chances. It will include an assessment of analysis and interpretation tools and the use of indicators and health profiles to evaluate progress, and will evaluate similarities in health policy and environmental policy with particular emphasis on progress made in the United Kingdom in response to global initiatives.

Environmental threats to health

Human actions that are detrimental to health – whatever the motivations – are clearly at odds with the notion of sustainability. We need look no further than chapter 6 of Agenda 21 and its commitment to protect and promote human health to see that this is the case (see below). Unfortunately, there are many complex and interacting environmental threats to human health arising from human economic development. Although it is not appropriate to review all of them here it will nevertheless be instructive to examine a few in more detail in terms of the spheres of environmental activity introduced in Chapter 1.

The sphere of production

Research on the links between health and work is strongly supportive of the notion that the work environment, through organisation and conditions of work and employment opportunities, has a significant effect on health (Kogi 1993). In favourable circumstances work can contribute directly to good health in terms of psychological well being – unemployment being a major cause of stress-related illness – and indirectly through the gains accruing from economic achievement (Shortt 1996; Graetz 1993). In less favourable circumstances, however, the workplace can be a determinant of ill-health with workers experiencing problems associated with indoor and work-created pollution and air quality: heat and humidity, noise, cigarette smoke, hazardous practices or dangerous substances, boredom, responsibility, and the like.

The sphere of consumption

Numerous health risks can be cited as being associated with the consumption of goods or the use of services such as water supply or electricity. Many of these can be packaged as 'lifestyle' factors – an unhealthy diet, the smoking of tobacco or the consumption of alcohol, or drugs use and misuse – and each include an element of 'consumer choice' (WHO Regional Office for Europe 1998). As in the sphere of production, risks in this group can have either a direct or indirect effect. If we take the case of food consumption the direct risks to health might include the nutrient deficiency arising from a poor diet, food contamination (perhaps introduced variously during cultivation, processing, storage, transportation or final food preparation), and over-consumption. Indirect risks to health include those specifically associated with the production processes designed to meet consumption demand – the use of pesticides to increase agricultural output, contamination of air and water from food manufacturing or processing plant, or atmospheric pollution arising from delivering food to the consumer – as well as those problems associated with the manufacture, and subsequent disposal, of packaging and waste.

The sphere of social capital/infrastructure

In viewing the links between health and sustainability it is the urban environment that has come under particular scrutiny (Hancock 1996; Harpham and Werna 1996; WHO Regional Office for Europe 1997). Partly this is an intrinsic fact of life, because globally the next century will see more than 50 per cent of the world's population living in urban areas. But as well, it is in the built environment that the sphere of social capital and infrastructure is most concentrated. Although increased urbanisation can have a positive affect upon health – it can for instance lead to much improved access to health services – there are also other, negative, aspects of the urban environment that can substantially affect health (Satterthwaite 1993) – housing, transport infrastructures, noise, energy consumption, waste disposal, crowding, and fear of crime amongst others. In these factors we can see health risks which might lead to hypothermia or respiratory illnesses (inadequate housing/heating), asthma (air pollution from motor traffic), injuries and accidents (crowding), diarrhoeal diseases (poorly managed household wastes), and others. Importantly as well, many physical characteristics of the urban environment can influence the incidence of mental disorders. In this, although there are chemical and physical factors that influence the bodies nervous system, it is the psychosocial factors – noise, overcrowding, poor services, levels of crime, social exclusion – that are most significant.

The sphere of 'nature'

Access to 'nature' – in broad term's open space and the countryside – represents an important dimension to the overall quality of life experienced by an individual. In particular, green spaces, the countryside, forests, and amenity landscapes provide a positive antidote to the stress-related factors found in the built environment (Parry-Jones 1990). Such areas can contribute directly to better health in facilitating physical activity – walking, running, or cycling – which has been shown to protect against cardiovascular disease and hypertension, improved weight control and reduction of obesity, an enhanced social life, and improved psychological well-being and stress management. Indirectly as well, nature contributes to the health of the human population through food supply and pharmaceuticals – species' extinctions irretrievably reduces the potential for new food sources or biomedical advances.

The sphere of physical sustainability

Any significant changes to background environmental processes – for the most part those operating at global scales – will likely have many impacts upon human health. Human-induced climate change, for instance would have direct effects through climatic hazards such as storms, floods and droughts and indirect effects through infectious disease transmission, impacts upon fisheries and agricultural production, and socio-economic disruption (WHO 1996). Similarly, stratospheric ozone depletion caused by the release of chloro- and bromofluorocarbons may lead to increased exposure to ultraviolet B radiation and the possibility of raised incidence of skin cancer and eye cataracts and suppression of the immune system. These and other global scale risks to health have been fully reviewed by Bentham (1994).

Geographical variations

Much of the evidence for the existence of an association between health, environment and economic conditions comes from studies of geographical inequalities. This is because all three domains demonstrate spatial variability, at a variety of different geographical scales, and similarities in their distributions can often be identified. At global scales the role of economic development in explaining international disparities in health is encompassed in the so-called demographic and epidemiological transition models. In this, countries experiencing low levels of economic development typically have relatively young populations with high fertility rates which make them susceptible to communicable diseases. This, coupled with the fact that infectious diseases are mostly poorly controlled in such countries, gives rise to high levels of mortality. As development proceeds, however, more effective controls over diseases begin to reduce mortality rates, particularly in the young, and population numbers rise as a result. In turn, fertility is likely to decline in response to increasing numbers and population growth is slowed down. Finally, as the population becomes less youthful the communicable diseases are replaced by non-communicable diseases as the dominant cause of mortality. Following these ideas health differences at the international scale are most commonly considered in relation to Gross National Product, with poorer, less developed, countries seemingly having worse health than richer ones. As such, health variations are deemed to be amenable to economic interventions.

Unfortunately the epidemiological transition models described above fail to account properly for the fact that similar variations in health have been shown to exist also at the infranational scale. In England and Wales, for instance, published figures at the regional scale indicate a strong mortality gradient from the North and West to the South and East – a long-standing imbalance which shows little sign of changing. The broad regional pattern is also reflected in the constituent counties, with the highest mortality rates occurring almost exclusively among counties in the North and West regions and Wales. Statistics for local authority districts, the lowest geographical level for which mortality data is routinely available in the UK, confirm the generally adverse mortality status of the declining urban industrial areas of the country (Britton 1990). In particular, the cities of Northern and Midland Britain, which have experienced recent losses of both jobs and population, are forming a new basis of concern about public health. Even within cities and towns there exist considerable internal geographical variations in health, with mortality varying substantially at the level of electoral wards. In Stoke-on-Trent, an industrial city in Midland England, for instance, the ward level mortality figures show significant variation around the overall Stoke value and range from 5 per cent to 69 per cent higher than the national experience (Kivell, Mason and Bennett 1991). Interestingly, at these more local scales it appears that even when areas of seemingly similar economic conditions are compared, there are often substantial differences in the observed health of the populations. Commonly this is attributed to micro-scale geographical variations in those same factors that are believed to be influential at broader scales – wealth and social position for instance – coupled with the tendency for people of particular social classes to concentrate in certain localities (Townsend and Davidson 1982).

Although social and socio-economic factors can be used to explain certain geographical inequalities, a number of studies have shown that people with comparable individual risk factors, perhaps through social class, have also shown differing health experiences according to where they live (Fox *et al.* 1984). Such findings suggest an environmental as a well as a social dimension to health inequalities. There are many environmental threats to human health including a large number, which might be described as 'naturally occurring', and many more which can be attributed more directly to human activities. Of the former we might include physical hazards such as landslides, earthquakes or floods; bio-logical pathogens and disease vectors; the non-availability of such natural resources as food, fuels or fresh water; or the many chemical agents present in the environment that are independent of human activities. Of course, even such nat-ural threats as these cannot be properly said to be without a human dimension because it is only after some human intervention that a natural environmental occurrence is transformed into an exposure and thus becomes a risk to health – the use of floodplains for housing for instance.

Of those other factors which are more directly attributable to human activities we can recognise two broad categories – 'traditional hazards' associated with lack of development, and 'modern hazards' associated with unsustainable develop-ment (WHO 1997). Traditional hazards to health can be closely linked with poverty and lack of sufficient development and might include the likes of inadequate household or community sanitation, uses of unsafe drinking-water, indoor air pollution arising from biomass fuel combustion for cooking and heat-ing, and unsatisfactory disposal of domestic wastes. Modern hazards, on the other hand, are related more to development and unsustainable consumption of resources and are exacerbated by absences of safeguards and policies for proper environmental control. Examples include water pollution from industry and intensive agriculture, urban air pollution from road vehicles and industrial emis-sions, climate change, acid rain, and stratospheric ozone depletion. Once again, the three-way interaction between health, environmental and economic factors is apparent – the 'risk transition', for example, records the change from 'traditional' to 'modern' that is often observed in countries undergoing economic development.

Assessing the relationship

The fact that environmental conditions often represent a risk to health has long been understood. Various pioneering studies in the nineteenth century, such as that by Chadwick (1842), gave rise to an understanding that poor living condi-tions were often the root cause of ill health. Other studies made comparisons of life expectancies in different parts of the country and illustrated that there was a strong geographical and environmental dimension to health. In one of the best known of these early investigations Dr John Snow used maps to trace the causes of deaths from cholera to an infected water pump in Soho (Snow 1855). Such studies contributed to the closing decades of the century becoming an important era of sanitary and housing reform by public authorities. Later, especially in the 1930s, studies of areal inequalities in health came to play a vital role in the realis-ation of the need for a National Health Service. More recently the aetiology of disease has been investigated with reference to air quality (Dunn and Kingham

1996); water quality (Pocock *et al.* 1980), waste incineration (Diggle *et al.* 1990), non-ionising radiation (Aldrich and Easterly 1987), soil contamination (Moir and Thornton 1989), industrial processes (Phillimore 1998), and others.

It is over-simplistic to attempt to explain geographical variations in health in terms of factors related solely to environmental or economic conditions but there are undoubted links which deserve attention. Unfortunately, establishing the relationship in any scientific sense is by no means straightforward. Part of the problem lies in the spatial complexity of environmental influences upon health. These may be naturally occurring or human-induced and may variously arise as controlled or inadvertent exposures from a wide range of localised point, line or diffuse sources. Specific pollutants will also respond in different ways to the par-ticular environmental medium to which they are introduced – water, land, air, etc. – and will possess varying propensities for transport away from their source. As a result, some health-related environmental hazards may be uniformly distributed over a wide area – sulphur dioxide emissions from power stations or nitrate pollu-tion of surface waters for instance – while others will show more restricted pat-terns which reflect only limited transport away from a local source – for instance traffic sources of nitrogen dioxide or exposure to electromagnetic fields (Nur-minen *et al.* 1996). In the same way, geographical patterns reflecting levels of economic development are similarly difficult to relate accurately to health inequal-ities. Even at the international scale where comparisons are perhaps the most reliable, total national income will not be the only 'economic' factor affecting health – the relative priority given to health care spending by different govern-ments, for instance, will influence health outcomes. Also it is by no means clear that economic progress will necessarily be accompanied by better health – increases in degenerative and pollution-related illnesses may be the outcomes of a more sedentary workforce and increased pollution arising from industrial growth and mechanisation.

The study of the environment and its links with health is inherently a spatial problem – environmental conditions vary geographically and health outcomes show spatial patterning as a consequence. For this reason, most of the methods that can be used to identify and measure the relationship have a strong geo-graphical basis. In policy terms too, these same spatial tools can be used to target preventative or remedial action or to monitor and assess the effects of such actions. One method that is commonly applied is that of *ecological analysis*. In this, relationships between health and environment are analysed on groups of people rather than on individual persons (Nurminen 1997) – the grouping variate usually being some geographical region. As such, the ecological approach has the advantage that it permits the study of very large populations and can often be undertaken from existent databases rather than needing specific individual-level data. Often, the ecological approach can be combined with *time series analysis* in which patterns in series of observations – such as mortality statistics – are investigated for evidence of causal relationships with environmental correlates. An alternative approach is to use *risk analysis*. This uses existing information and understanding of the environment-health relationship, usually based upon previous studies in other areas, to estimate the potential health effects of exposure to a particular hazard. Unlike ecological and time series analysis the application of risk analysis does not require the existence of health data for the area under study – the risk to health being inferred from data on environmental exposures.

For any method to be applied properly there must be data available to describe the situation in the area under investigation. Here, the notion of *indicators* becomes important. Indicators go beyond the raw data derived from simple measurement. Instead, they represent a re-expression of the raw data by converting them into information to support the process of decision making – a transformation which gives the data 'added value'. Given the interactions between health, environment and economy that we have already identified we can see that in the field of environmental health contributory indicators can be drawn from a number of separate disciplines. Thus for instance, we may utilise economic indicators (such as GDP or unemployment rate) or social indicators (e.g. deprivation or levels of crime), as well as those more directly describing the physical environment (e.g. levels of ionising radiation or surface water quality) or human health (e.g. levels of morbidity or mortality). More usefully still, efforts are now being made to develop indicators that express the links between the associated fields. Importantly, the policy of sustainable development comes into play here as it is the need to describe and monitor progress towards sustainability that has become the driving force towards the development of new 'integrative' indicators – and of course health issues figure prominently in these developments. *The Indicators of Sustainable Development for the United Kingdom* (DoE 1996), for instance, has health identified prominently in several of its broad aims: 'A healthy economy should be maintained to promote quality of life while at the same time protecting human health and the environment' and, 'Damage to the carrying capacity of the environment and the risks to human health and biodiversity from the affects of human activity should be minimised'.

Similar efforts to develop core indicators at international level (e.g. OECD 1993; UNEP/RIVM 1994) and local level (e.g. Local Government Management Board 1994) also stress the centrality of health in issues of sustainability. Unfortunately, these core sets of indicators are sometimes rather too broad to properly identify the specific environmental-health risk that concerns us here. Instead, *environmental health indicators* (Corvalan *et al.* 1996) are required that specifically provide a measure of the effect upon health of an exposure to an environmental hazard. Here, the WHO has been particularly active both in terms of promoting environmental epidemiology and the need for reliable environmental health indicators (WHO 1983) and through implementation of supranational programmes, which require the use of such measures (WHO 1992).

WHO Healthy Cities project

The need for an integrative approach to health policy and planning and the importance to this of information is exemplified by the Healthy Cities project, which emerged from a World Health Organisation meeting in Lisbon in 1986.

The Healthy Cities project has been fully discussed elsewhere (Ashton 1992), but a number of the resolutions are particularly relevant to note here. Of prime importance is the intention that participating cities should: 'carry out a community diagnosis for the city, down to the small area level, with an emphasis on inequalities in health and the integration of data from a variety of sources including the public perception of their communities and their public health' (Ashton 1992: 9). In this, we can see the obvious importance of information on the existing health and social and environmental conditions within cities and the realisation that differences in health will exist even within a relatively confined geographical region. We can also see the direct influence of Agenda 21 which advocates surveying local conditions as a prerequisite to developing local health plans.

The Healthy Cities project has advocated the use of *city health profiles* as an important tool in creating environments supportive of sustainable health, often being used as the basis of a health plan to improve the health of the city population. A health profile will be concerned with providing measures of human health alongside information describing various aspects of the physical environment, measures of deprivation, and levels of health-damaging or health-promoting behaviour (WHO Regional Office for Europe 1995). Usefully, we can examine the typical contents of a health profile to again illustrate the existence of multiple environmental impacts upon health – and the need for inter-sectoral action to exercise control over their effects. As a fundamental prerequisite, information is usually included on the size of the population and its demographic description in terms of gender and age structures – here the national population census will be a major source of information. Basic statistics on births and deaths – the so-called 'vital statistics' – will similarly be incorporated from central, usually government, sources. Such information is so fundamental and readily available that it would be expected in all health profiles. Other data might be less accessible through reasons of cost or confidentiality and might not be as easily incorporated. For instance, useful indicators of health status itself are not always readily available. In fact, a common approach is to use mortality as a surrogate measure of health – data on morbidity being harder to come by. Mortality data will usually be presented as ratios (commonly compared with national figures) for selected causes of deaths, or as 'avoidable deaths' (deaths from conditions where there are effective forms of prevention or treatment), or as 'years of life lost' (the difference in years between age at death from specific causes and national life expectancy). Morbidity data which may be available might include hospital admissions and primary care attendance's by cause and various statistics from statutory procedures such as cancer registration or notification of infectious diseases. The inclusion of data on the various determinants of health will be guided by epidemiological evidence. Adequate housing, for instance, is now recognised as an important prerequisite for good health and the health profile would cover, wherever possible, aspects of homelessness, housing quality and amenities, tenure, density of occupation, and the like. The influence of socio-economic factors upon health would also be reflected in the profile. Topics for inclusion might cover education, income, employment, crime levels, etc. Similarly, the physical environment will impact upon health and statistics would be collected on such topics as air and water quality, public open space, noise, traffic densities, water and sewage services, infestations, etc. Lifestyle measures will also be guided by evidence of a relationship to health. Thus levels of smoking – either measured directly by population survey, or indirectly through hospital admissions for smoking-related diseases – alcohol consumption, misuse of drugs, exercise, and diet will all be incorporated in the profile if they are available. Finally data on the physical and social infrastructure of the city will be needed. This will include the likes of transport and telecommunications facilities, city planning, urban renewal programmes, and leisure facilities, as well as the existence of public health policies and services such as immunisation, screening, family planning, health education or drugs misuse programmes, and levels of access to all types of primary and secondary health care facilities. Initially, the Healthy Cities project adopted a set of 53 such indicators to assist participating cities in gathering appropriate data to describe health in their locality – these have since been reduced to a more concise set of 32 indicators grouped as representing health, health service provision, and environmental and socio-economic conditions. The Healthy Cities movement demonstrates a specific response to a local, mainly urban, problem but current environmental debates point, as well, to risks to human health such as global warming and the destruction of the ozone layer which require policy interventions on a much wider scale. An interesting debate here is the way solutions to a local problem can achieve global benefits as well.

A sustainable approach?

The Healthy Cities project provides us with an example of intersectoral policy geared towards achieving sustainable health in an urban setting – we will return to look at other policy responses to sustainable development below. Implicit in the Healthy Cities approach is the notion that a concern for health and similar concerns for environmental quality and economic development go hand in hand, and that there is much to gain from a close working relationship between the different sectors operating within the urban arena. In turn, our use of sustainability as an integrative concept which brings together the interests of all sectors of society implies that sustainability and health – as with other sectors – have inextricable links. Hancock (1996) provides some examples that we can use here. One obvious link between health and sustainability concerns the issue of traffic and urban design. We have seen already that traffic is a major source of air pollution and that air pollution can lead to respiratory disease or can trigger asthma attacks. Reducing the use of cars, improving public transport, developing cleaner fuels or more efficient engines, and creating environments which encourage pedestrians and cyclists – all obvious sustainability objectives – will also be creative of better health, not just through improvements in air quality, but also because of improved opportunities for exercise and the creation of patterns of travel which are likely to be less stressful and demanding on the individual. Other examples given by Hancock include the following: the use of ecological or 'living' machine technologies for sewage treatment; local food production to replace energy inefficient transport of foods over vast distances; and a switch to low-meat, high-vegetable diets that enable a more environmentally sustainable system of agriculture.

Unfortunately, although there are many benefits to be gained from efforts to improve sustainability, we must also realise the potential for conflict. In broad terms, for instance, we can recognise an apparent conflict between the protection of health and the pursuit of wealth – the latter with its attendant impacts upon the environment. A simple everyday example of this type of conflict is the trade-off that is needed to balance any exposures to health-related hazards in the workplace – our 'sphere of production' – against the improved quality of life that usually comes with regular employment. We have seen already that organisation and conditions of work have a profound effect on health – the sustainable solution to the problem is to maintain levels of employment opportunities while at the same time encouraging good practice in workplace conditions and organisational development. Action of this kind can of course be self-perpetuating: improved work conditions will lead to a healthier work force which in turn will encourage improved productivity, and will ultimately provide further opportunity for a still healthier and yet more productive workplace. A second type of conflict, that between local and global solutions to the environment-health problem, also requires addressing. For an example of this type of conflict we might return to our traffic problem to see how action that is beneficial at the local level can have an opposite, adverse, health effect on a global scale. One way to reduce exposures to local traffic-related pollution is to develop out of town shopping in order to spread the pollutant load. Such policy is likely to improve the local situation by reducing 'hot spot' concentrations but will actually exacerbate the global air pollution problem by creating extra car journeys to more distant sites. A sustainable solution here, one that reconciles the local with the global, is to *reduce* local car traffic rather than just *displacing* it. This could be achieved through use of the planning system to reduce the need to travel or by increasing the availability and attractiveness of public transport (Chartered Institute of Environmental Health 1995).

Policy frameworks

We have argued that economic and other forms of human activity have numerous impacts upon the local and global environment and that hazards exist as a consequence of this that represent direct or indirect risks to human health. We have also seen that good health is essential for social and economic development and is fundamental to sustainable growth. It is clear then that the agenda for sustainable development has much in common with that for public health. In the same way that the public health agenda has evolved to embrace a very wide ranging set of interacting social, economic and environmental factors, so too has that for sustainable development. The latter is now much less focused upon economic development *per se* and is more concerned with human development, with economic development being seen as one way, but not the only way, of achieving it. The principal international frameworks that exist for health and sustainable development – the World Health Organisation's 'Health for All 2000' (HFA 2000) and the United Nation's Agenda 21 are thus inextricably interconnected.

HFA 2000 is the WHO global strategy for health development and has origins that can be traced back to the Stockholm Conference on Human Environment in 1972 and the WHO 30th Assembly in 1977. It was here that the overall target was set to attain for all citizens a level of health by the year 2000 that will permit them to lead 'a socially and economically productive life'. Individual targets are strongly directed towards social and environmental determinants of health with the premise that implementations of HFA 2000 requires intersectoral action on lifestyle and environment as well as on health service delivery. Specific social targets include the promotion and support of healthy living, reductions in health-damaging consumption of tobacco and alcohol, and improved intersectoral policies for health education. Environmental targets are aimed at providing social and physical environments supportive to health, including improvements in the workplace. Specifically, HFA 2000 requires that there should be access for all to safe drinking water and neither water nor air pollution should present a threat to health. Similarly, risks to health from waste or soil pollution should be properly controlled and there should be improved food quality and safety. Importantly, as well, HFA 2000 sets specific targets that look for the development of policy on environment and health for sustainable development.

Implementations of HFA 2000 at national level vary from those countries that have adopted the strategy fairly comprehensively to those whose governments have adopted a more selective strategy. In the United Kingdom, the government's policy document *The Health of the Nation* (Department of Health 1992) outlined the need to identify the main health problems and focussed upon setting targets for improvement in five selected causes of ill-health and death. In setting out its targets the document makes it clear that health is determined by a range of influences, both personal (e.g. genetic inheritance, personal behaviour) and public (e.g. physical environment and social environment). In the latter context, the role of a number of government departments was highlighted, including that of the then Department of the Environment which had responsibility for, *inter alia*, environmental protection, water quality, housing and, through inner-city programmes, for providing some funding for health promotion and primary health care. Unfortunately, although the Health of the Nation Document clearly recognises that multiple factors interact to shape the health of the population, the strategy it proposes is essentially disease oriented and confines itself to a focus on lifestyle factors in order to improve health (Gustaffson 1997). A new agenda for public health in the UK, adopting a strategy which highlights the equal importance of environmental, social and economic factors, was published in 1998 as the Government's consultative Green Paper *Our Healthier Nation*. This provides further explicit recognition by Government that a link exists between heath and a wide range of contributory factors and that various national policies such as social exclusion, welfare to work, education, transport and air quality impact directly on health outcomes.

Agenda 21 – the United Nations programme of action on sustainable development – was agreed at the UN Earth Summit in 1992 and sets out how countries can work together towards sustainable development. Its local implementation, Local Agenda 21, is concerned with achieving action at community level by adopting a partnership approach. The framework that is laid down by Agenda 21 emphasises very clearly that health is indispensable to sustainable development and therefore should be a primary concern in formulation of policy for economic and social development. The principal focus for this concern for health is chapter 6 – 'Protecting and promoting human health'. Chapter 6 highlights several specific health challenges including the particular problems of urban health, reducing the risks from environmental hazards, protecting vulnerable groups, controlling communicable diseases and meeting primary health care needs. In addition, Agenda 21 addresses a wide range of other health issues ranging from occupational health and safety through to the health impact of atmospheric changes. In all of these the need for intersectoral action involving health is prominent. The UK government response to its commitment to Agenda 21 was to publish its strategy document *Sustainable Development: The UK Strategy* (HM Government 1994) which had the Brundtland Report definition of sustainable development at its core. The general principles of the UK strategy encouraged a shared responsibility between government and others and integration of environmental policy into all aspects of government. However, the overall tone suggested that protection of the environment should not stand in the way of economic competitiveness. More recently, the current Labour Government has sought to rectify the preoccupation evident in the original strategy of reconciling the needs of wealth creation with those of environmental protection. Instead, the perspective on sustainable development has been broadened to incorporate social objectives – including health – as well as economic and environmental intentions.

It is apparent that although the HFA 2000 strategy was originally developed to focus upon health and the Agenda 21 strategy to focus on environment, there are very obvious similarities between the two frameworks. Not only is there a shared United Nations context and parallel requirements for action at both global and national levels, there is also clear commonality of principle – the desire for intersectoral collaboration, community participation and empowerment, social justice and reduction of inequalities and sustainability – and a shared agenda incorporating social, environmental and economic issues as they impinge upon overall quality of life. In addition, we can see further evidence of a convergence of interest between health and the environment in the WHO European Charter on Environment and Health and the establishment of the WHO European Centre for the Environment and Health. At national level as well, the evolving links between health and sustainable development is increasingly apparent. In the UK, for example, a number of key agencies have begun to explore in recent years some of the interconnections between health, environment and sustainability by such means as themed conferences and specialist publications on the subject. Dooris (1998) identifies several such developments, including the following:

- the establishment of the UK Health for All Network working group on Environment and Health (1989) and its 1994 conference theme 'Health and the Environment: Towards Sustainable and Support Environments for HFA';
- publication in 1995 of the Public Health Trust's 'Sustainable Development and Health';
- publication in 1996 of the UK Government National Environmental Health Action Plan;
- publication in 1997 of the Local Government Management Board Round Table Guidance Notes on Health and LA21 and the Association for Public Health manifesto 'Sustaining the Public Health'; and
- publication in 1998 of the Government's Green Paper 'Our Healthier Nation' which emphasises the need for links to be developed with Local Agenda 21 and Environmental Health Action Plans.

Conclusion

The Rio Declaration's first principle that 'Human beings are at the centre of concerns for sustainable development. They are entitled to a healthy and productive life in harmony with nature' puts health firmly on the environment and development agenda. We began this chapter on health and sustainability by examining just a few of the possible environmental threats to health, linking them to the five 'spheres of environmental activity' that were identified in the introductory chapter. We saw that there are a variety of techniques that can be used to measure or predict the strength or significance to health of many of these environmental hazards, and that the threat to health that they pose has an important geographical dimension. From studies employing such techniques it is apparent that where you live can substantially influence your chances of good health. The fact that many of these threats are the so-called 'modern hazards' that are related to development and consumption of resources suggests that environmental health represents a key challenge to policies for sustainable development. We have seen how global initiatives such as Agenda 21 and Health for All 2000, and actions directed more at local populations such as the Healthy Cities project, are beginning to address the problem through adopting integrative strategies aimed at promoting intersectoral co-operation on health promotion and protection. Such co-operation and co-ordination of effort between different sectors will be vital if the complex interaction of factors affecting health is to be properly addressed. The importance of baseline information and the development of indicators as a measure of progress towards sustainability is acknowledged here. Finally, we have seen how a convergence is taking place between those agendas that have been set up primarily to focus upon health and those established with a primary focus upon the environment.

Undoubtedly there are signs that the inter-linked problems of health, environment and economic development are being addressed – particularly in the global and national infrastructures that are now being put in place. However, there is still much to be done. More evidence is required on specific disease–environment relationships and better systems for measuring and monitoring exposures need to be put in place before the risk to health can be properly assessed at different geographical scales. We also need to recognise some of the wider dimensions of health and particularly to develop better measures of human well-being – physical and psychological – to supplement the existing reliance upon indices for death and disease. There also needs to be more practical guidance on how to link sustainability and health and how this can be translated into policy making – particularly at the local level where microenvironmental factors related to living and working conditions will likely add further complexity to existing macroenvironmental-health relationships. The quality of the environment is one of the principal causes of differences in people's health. The challenge is for the progress towards sustainability to sufficiently impact upon environmental improvements to substantially reduce the health inequalities that are apparent today.

References

Aldrich, T. and Easterly, C. (1987) Electromagnetic fields and public health. *Environmental Health Perspectives,* 75, pp. 159–71.

Ashton, J. (1992) *Healthy Cities.* Milton Keynes: Open University Press.

Bentham, G. (1994) Global environmental change and health. In D.R. Phillips and Y. Verhasselt (eds) *Health and Development.* London: Routledge, pp. 33–49.

Britton, M. (1990) Geographic variation in mortality since 1920 for selected causes. In M. Britton (ed.), *Mortality and Geography: A Review in the Mid-1980s, England and Wales.* London: HMSO.

Chadwick, E. (1842) *Report from the Poor Law commissioners on an enquiry on the sanitary conditions of the labouring population of Great Britain.* London: HMSO (reprinted 1965).

Chartered Institute for Environmental Health (1995) *Environmental Health for Sustainable Development.* London: CIEH.

Corvalan, C., Briggs, D. and Kjellstrom, T. (1996) Development of environmental health indicators. In D. Briggs, C. Corvalan, and M. Nurminen, *Linkage Methods for Environment and Health Analysis: General Guidelines.* Geneva: WHO.

Department of Health (1992) *The Health of the Nation: a strategy for health in England.* London, HMSO.

Diggle, P.J., Gatrell, A.C. and Lovett, A.A. (1990) Modelling the prevalence of cancer of the larynx in part of Lancashire: a new methodology for spatial epidemiology. In R.W. Thomas (ed.) *Spatial Epidemiology.* London: Pion, pp. 35–47.

DoE (1996) *Indicators of Sustainable Development for the United Kingdom.* London, HMSO.

Dooris, M. (1998) *Health and Local Agenda 21: Towards Integrated National Networking in the United Kingdom.* Report for Healthy Cities National Networking Consultation Meeting, Rotterdam, 1998.

Dunn, C. and Kingham, S. (1996) Establishing links between air quality and health: searching for the impossible? *Social Science and Medicine*, 42(6), pp. 831–41.

Fox, A., Jones, D. and Goldblatt, P. (1984) Approaches to studying the effect of socio-economic circumstances on geographic differences in mortality in England and Wales. *British Medical Bulletin*, 40(4), pp. 309–14.

Graetz, B. (1993) Health consequences of employment and unemployment: longitudinal evidence for young men and women. *Social Science and Medicine*, 36(6), pp. 715–24.

Gustaffson, U. (1997) Improving the nation's health: the relevance of place in English health policy. *Health and Place*, 3(4), pp. 281–4.

Hancock, T. (1996) Health and sustainability in the urban environment. *Environmental Impact Assessment Review*, 16, pp. 259–77.

Harpham, T. and Werna, E. (1996) Sustainable urban health in developing countries. *Habitat International*, 20(3), pp. 421–29.

HM Government (1994) *Sustainable Development: The UK Strategy*, Cm2426. London: HMSO.

Kivell, P.T., Mason, K.T., and Bennett, G. (1991) *A Health Profile of North Staffordshire*. Unpublished report to the North Staffordshire Health Authority.

Kogi, K. (1993) Workplace strategies for the control of work-related risks. *Environmental Research*, 63(1), pp. 88–94.

Labont, R. (1991) Econology: Integrating health and sustainable development. Part 1: Theory and background. *Health Promotion International*, 6(1), pp. 49–65.

Local Government Management Board (1994) *Sustainability Indicators – Guidance to Pilot Authorities*. London: Touche Ross.

Moir, A.M. and Thornton, I. (1989) Lead and cadmium in urban allotment and garden soils and vegetables in the United Kingdom. *Environmental Geochemistry and Health*, 11, pp. 113–19.

Nurminen, T., Nurminen, M., Corvalan, C., and Briggs, D. (1996) Exposure assessment. In D. Briggs, C. Corvalan, and M. Nurminen, *Linkage Methods for Environment and Health Analysis: General Guidelines*. Geneva: WHO.

Phillimore, P. (1998) Uncertainty, reassurance and pollution: the politics of epidemiology in Teesside. *Health and Place*, 4(3), pp. 203–12.

OECD (1993) OECD core set of indicators for environmental performance reviews. *Environmental Monograph No 83*. Paris: OECD.

Parry-Jones, W.L. (1990) Natural landscape, psychological well-being and mental health. *Landscape Research*, 15(2), pp. 7–11.

Pocock, S.J. (1980) British regional heart study: geographic variations in cardiovascular mortality and the role of water quality. *British Medical Journal*, 280, pp. 1243–49.

Satterthwaite, D. (1993) The impact on health of urban environments. *Environment and Urbanization*, 5(2), pp. 87–111.

Shortt, S.E.D. (1996) Is unemployment pathogenic? A review of current concepts with lessons for policy planners. *International Journal of Health Services*, 26(3), pp. 569–89.

Snow, J. (1855) *On the Mode of Communication of Cholera*. 2nd edition. London.

Townsend, P. and Davidson, N. (1982) *Inequalities in Health: The Black Report*. Harmondsworth: Penguin.

UNEP/RIVM (1994) *An overview of Environmental Indicators: State of the Art and Perspectives*.

WHO (1983) *Environmental Health Criteria 27. Guidelines on studies in environmental epidemiology.* Geneva: WHO.

WHO (1992) *Twenty Steps for Developing a Healthy Cities Project.* Copenhagen: WHO.

WHO (1996) *Climate Change and Human Health.* Geneva: WHO.

WHO (1997) *Health and Environment in Sustainable Development: Five Years after the Earth Summit.* Geneva: WHO.

WHO Regional Office for Europe (1995) *City Health Profiles. How to Report on Health in your City.* Copenhagen: WHO.

WHO Regional Office for Europe (1997) *Sustainable Development and Health: Concepts, Principles and Frameworks for Action for European Cities and Towns.* Copenhagen: WHO.

WHO Regional Office for Europe (1998) *Social Determinants of Health: The Solid Facts.* Copenhagen: WHO.

World Bank (1993) *World Development Report 1993: Investing in Health.* Oxford: Oxford University Press.

Part 2

Historical perspectives on sustainable livelihoods

6 Population, food and agriculture in mid-nineteenth century England

A. D. M. Phillips

Introduction

The growth in the demand for food in England and Wales during the nineteenth century was spectacular. The size of the food market may be regarded as a function of population numbers and per capita demand for food, and both variables underwent considerable change over the century. In terms of numbers alone, population in England and Wales increased from 8.9 to 32.5 million between 1801 and 1901, and this rise in itself, if consumption levels remained uniform, would have enlarged the market 3.7 times. Per capita demand was largely dependent on wage levels, and working on the assumption that the proportion of income devoted to food was constant – probably an underestimation of the importance of food in improving family budgets – any rise in real earnings would enlarge the market for food. Over the century the rate of growth in real wages – conservatively estimated at 4.4 times – would have amplified the demand for food stemming from straightforward population increase. By combining the rates of growth of population and real wages, some overall indication of the scale of increase in the size of the food market may be obtained. This measure – the potential demand for food – has been calculated at rising at the astonishing rate of 15.6 times between 1801 and 1901 (Walton 1990).

At the same time as the food market grew, its composition was subject to marked change. Arising from the general improvement in real income, diets began to be modified as funds were available not only for the staple items of consumption, largely grains, but increasingly for more varied and more satisfying foods. As a contemporary commentator could note in 1880: 'Thirty years ago, probably not more than one-third of the people of this country consumed animal food more than once a week. Now, nearly all of them eat it, in meat, or cheese, or butter, once a day. This has more than doubled the average consumption per head; and when the increase of population is considered, has probably trebled the total consumption of animal food in this country' (Caird 1880). The impact of such varying trends in consumption on the food market may best be demonstrated in the respective price movements of agricultural produce over the century. Thus, the prices of grains, especially wheat, were characterised by periods of decline and at best of stability, a contrast to those of animal produce which were normally buoyant and at worst constant.

For the domestic agricultural industry, whose principal product was food, these developments heralded challenges on an unprecedented scale. To be able to approach, let alone sustain, these ever growing demands for more food in general and for more animal produce inn particular, necessitated drastic revaluation of both production techniques and farming systems. Such reassessment would have been exacerbated by the fact that the limits of cultivation in England and Wales had been reached by the early decades of the nineteenth century and thereafter receded: significant increases in agricultural output for most of the century were not achievable by expanding the cultivated area. Only the removal of protection from domestic farm produce, beginning with the repeal of the Corn Laws in 1846 and allowing food products from overseas to be freely imported, would have lessened the pressure to rationalise agricultural practices (Grigg 1989). The extent to which, and the manner in which, agriculture in England and Wales responded to these patterns of demand over the course of the nineteenth century cannot at present be determined with precision. If the broad canvas remains to be completed, the nature and problems of the relationship, with its attendant implications for distributions of land use and cropping, rates of technological development, structures of farming systems, and levels of farming prosperity, may be illustrated at a smaller scale. To that end the present study analyses agricultural production and food supply in Staffordshire between 1840 and 1870.

The middle decades of the nineteenth century have long been regarded as a period of prosperity for farming; indeed the years between 1850 and 1870 have been termed the golden age of English agriculture (Ernle 1961). The basis of that prosperity has been attributed by both contemporary and present-day commentators to structural changes in farming systems and unparalleled technical progress (Caird 1880; Jones 1968). Against the background of the long-term decline in wheat prices after 1815 and the immediate threat of loss of protection on cereals through the repeal of the Corn Laws, agriculturalists at mid-century at both national and local levels questioned the ability of current farming systems and management practices to be financially viable. Aware of the overall expansion of the food market and of the changing pattern of food consumption, they argued that improvement in prosperity levels required the application of new agricultural strategies that could best take advantage of such trends, particularly those that increased output of agricultural products, both cereal and livestock, and those that encouraged a shift from grain to livestock in mixed farming systems (Caird 1852; Ford 1846).

If the agricultural community were persuaded by such arguments, the structural and output changes in farming systems, essential to benefit from the predicted movements in demand and prices, could be achieved only through the large-scale adoption of a series of improvements in agricultural techniques. These encompassed the enlargement of the stock and range of farm buildings, the effective underdraining of farmland, the increased application of fertilisers and feedstuffs, the acceptance of new forms of mechanisation, and the employment of more progressive methods of cultivation and livestock management. By intensifying farming methods, these technical developments – individually and collectively – were capable of making possible high levels of production. However, while they offered the prospect of greater flexibility and technical efficiency in farming, described generally as high farming, their adoption involved major capital outlay and the application of the most advanced knowledge on the part of landlords and tenant farmers (Collins 1995; Thompson 1968; 1981).

The trends in food demand in Staffordshire between 1840 and 1870 were such as to give every incentive for the ready implementation of these new farming strategies. Total population rose from 508,000 in 1841 to 857,000 in 1871, an increase of 1.7 times. Over the same period most wage indices point to a growth in real incomes for all workers in the county. Thus, the nominal weekly wage of agricultural labourers rose 1.6 times from 9s around 1840 to 14s in 1867–70. Applying this rate of income growth to population expansion yields an increase in the potential demand for food in Staffordshire of 2.6 times over the three decades at an annual rate of increase of 3.2 per cent. However, this level of increase should be regarded as a minimum indication of the growth in the food market. At mid-century, a differential of at least 30 per cent has been detected in wage levels between urban and rural employment in favour of the former. As the populations of the urban-industrial complexes of the Black Country and the Potteries were expanding at a rate faster than the county's overall population (a two-fold increase between 1841 and 1871, representing at the latter date 78 per cent of Staffordshire's total population), the pressure of this differential would have created even greater growth in real wages and by extension in the potential demand for food (Phillips 1985; 1997).

An indication of the type of food in demand from this growing and increasingly wealthy population can be obtained from an examination of the price movement of the major agricultural products in Staffordshire markets (Table 6.1). Distinct and diverging trends were evident in the prices of grain and livestock products. The price of wheat, although subject to much fluctuation, fell gently over the period, being 7 per cent lower on average in 1864–69 than in 1840–44, and

Table 6.1 Price indices of butter, beef, mutton and wheat in Staffordshire markets, 1840–70 (average price of 1840–44 = 100)

Product	1840–44	1845–49	1850–54	1855–59	1860–64	1865–69
Butler[a]	100	113	96	110	118	124
Beef[b]	100	102	97	118	120	137
Mutton[c]	100	110	106	128	131	133
Wheat[d]	100	90	73	100	86	93

Source: Sturgess 1965.

Notes
[a] Average price of butter, 1840–44 = 12.7d per lb.
[b] Average price of beef, 1840–44 = 5.95d per lb.
[c] Average price of mutton, 1840–44 = 6.43d per lb.
[d] Average price of wheat, 1840–44 = 57.76s per imperial quarter.

suggesting little shift in demand and by association consumption levels. Against this, with the exception of the period 1850–54, the prices of butter, beef and mutton pursued an upward course. They were respectively 24, 37 and 33 per cent higher in the last quinquennium than in the first, indicating not only the increased demand for these products but also the potential profitability afforded by the adoption of livestock enterprises. Although Staffordshire, like all other parts of England and Wales, should not be regarded as a closed system for food production and consumption, the adoption of the structural and technical changes propounded by contemporary agricultural writers represented not only a way of meeting and sustaining the demand for food generated in the county but at the same time promised a significant improvement in prosperity, well beyond the capabilities of existing farming systems.

Although infrequent and often flawed, data are available to undertake some assessment of the responses in Staffordshire agricultural to these changing patterns of demand. Indications of adjustments in farming systems may be obtained from an examination of land use and cropping trends. While official agricultural statistics detailing land use and cropping began to be collected in England and Wales on an annual basis from 1866, before that date comprehensive cover is lacking. For Staffordshire the most complete record of land use and cropping is found in the tithe surveys of around 1840. From the preambles to tithe awards and agreements and from tithe files, data identifying major agricultural land uses for 83 per cent and individual crops for 45 per cent of the 1870 cultivated area of the county are extant. Although partial and estimated, these figures when compared with the agricultural returns of 1870 may be used to review the nature and extent of land use and cropping change in the county between 1840 and 1870.

Around 1840, in all parts of the county, various forms of mixed farming were practised (Figure 6.1). Overall, permanent grassland was already more significant than arable. In the northern parts of the county, on heavier and more upland soils, this dominance was clearly marked, with dairying being the most widespread livestock enterprise. On the generally lighter soils in the south, the area devoted to arable increased, and the livestock enterprises developed in this part were a mix of dairying, cattle breeding and sheep. On arable throughout the county wheat was normally the dominant cereal (Phillips 1973). With the spatial dominance of grassland, farming systems in Staffordshire were so structured around 1840 to be in a position to adapt readily to trends in food demand.

By 1870 an element of change can be identified in farming systems. The pattern of major agricultural land uses had become more rationalised, with the respective dominance of permanent grassland and arable in the north and south of the county being clearly evident (Figure 6.1). The decline in the proportion of the cultivated area given over to bare fallow, thus increasing the cropped acreage, indicates an attempt to expand overall agricultural output (Tables 6.2 and 6.3). Some effort had been made to extend livestock enterprises by increasing the area of crops devoted to animal usage. Thus the proportion of the cultivated area in permanent grassland rose, with a corresponding fall in the arable acreage. However, the scale of the movement was slight, a growth of 10 per cent between 1840

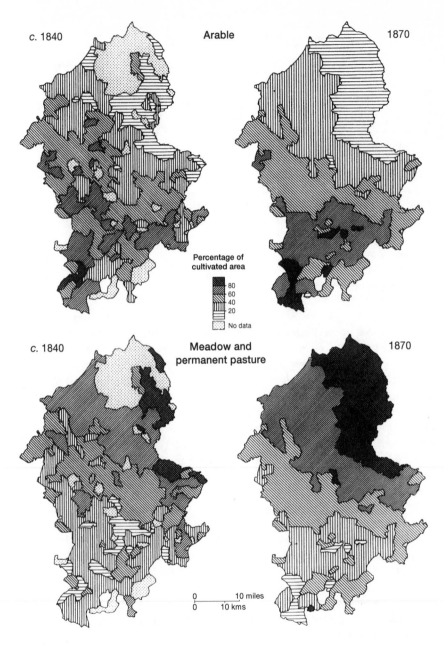

Figure 6.1 Agricultural land use in Staffordshire, *c.* 1840–70

Source: Phillips 1997

Table 6.2 Agricultural land use in Staffordshire, 1840–70

Date	Recorded acreage	Percentage in	
		Arable (%)	Meadow and permanent pasture (%)
c. 1840	474,393	46.1	53.9
1870	574,420	40.8	59.2

Source: Phillips 1997.

Table 6.3 Cropping in Staffordshire, 1840–70

Date	Recorded acreage	Cereals (%)	Wheat (%)	Barley (%)	Oats (%)	Pulses (%)	Green crops (%)	Rotation grass (%)	Fallow (%)
c. 1840	260,387	22.9	10.3	6.0	6.4	1.0	5.3	12.4	4.3
1870	574,420	20.7	9.4	5.7	5.4	1.4	7.7	9.3	1.6

Source: Phillips 1997.

and 1870, well below the levels of increase in the demand for food and the prices of animal products. At the same time the combined area of green crops and rotation grasses – further sources of animal fodder – demonstrated little expansion, remaining almost stable at about 17 per cent of the cultivated acreage. In 1870 the proportion of the cultivated area in cereals had fallen, but only slightly, less than 10 per cent. Although the acreage planted to wheat had declined by about 9 per cent, broadly in keeping with the overall movement of its price, it still remained the major cereal throughout the county. Overall, in terms of agricultural land use and cropping, a significant expansion in the area of those crops associated with the development of livestock enterprises is difficult to determine in the county.

Of course, a growth in livestock activities could be achieved without drastic changes in cropping. While broadly maintaining current cropping regimes, livestock output could be increased by means of purchasing artificial feedstuffs and by using on-farm-produced grains, especially wheat, as feed. The evidence for such advances in productivity should be sought in both numbers of livestock and output per animal. Reliable information on the latter is difficult to provide, and data on the former are little better. Records of livestock numbers for the county are not available until the introduction of official agricultural statistics in 1866. If examination of livestock numbers at the county level is precluded, data are extant for parts of Staffordshire over the period, relating to dairy cattle. Although fattening cattle and sheep were recognised as other animal enterprises in Staffordshire, dairying according to Henry Evershed, author of the Royal Agricultural Society of England's prize essay on the county's farming, had come to be the dominant livestock activity by 1869 (Evershed 1869). While partial, these data should provide some indication of the performance of the major livestock concern in the county.

For eight parishes in east and central Staffordshire, the tithe files record the number of diary cows around 1840. The official agricultural statistics did not distinguish dairy cows as a group but listed cattle numbers between cows and heifers in milk or in calf, and other cattle. If cows and heifers in milk or in calf can be equated with dairy cows in the tithe files, comparable data may be provided for 1870 for the same eight parishes from the parish summaries of the official agricultural statistics (Table 6.4). Aggregating the figures, the results would suggest that while the head of dairy cattle increased in these parishes between 1840 and 1870 it was of a limited order. The density of dairy cattle per 100 cultivated acres in 1870 was 14 per cent greater than around 1840. As the 1870 density was not far removed from the then Staffordshire average of 11.6 dairy cattle per 100 acres, the changes in this group of parishes may be seen as broadly representative of developments in the county as a whole. While recognizing that no account has been made of yield per beast, the data from these sample parishes are difficult to reconcile with a marked expansion of dairying, the predominant livestock enterprise in the county, and reveal a rate of growth well below the increases in the prices of animal products.

While available evidence points to small improvements in livestock productivity, that for cereals suggests a more marked development. Data on cereal productivity relate predominantly to wheat, the dominant grain crop. Contemporary estimates of wheat yields in Staffordshire over the three decades rise from 21.3 bushels per acre around 1840, to 28 bushels in 1850, to 28.5 bushels in 1861, and 29.5 bushels in 1870. If reliable, these estimates indicate an overall rise in per acre output of 38 per cent over the period. Some confirmation that these isolated estimates may be regarded as representative of the general trend in wheat output per acre in the period is provided by the only extant run of wheat yields for the county based on the crop-cutting data of a firm of Liverpool corn merchants between 1821 and 1859. While the absolute values in the series must be questioned it indicates that average wheat yields in Staffordshire rose by 34 per cent between 1839–43 and 1855–59 (Phillips 1997). Despite the great emphasis of contemporary agriculturalists on the benefits of expanding livestock enterprises and the limited financial opportunities offered by cereals, the highest rates of growth in agricultural output in the county were achieved in the production of wheat. Nevertheless, the rate of increase in

Table 6.4 Density of dairy cattle in eight Staffordshire parishes, 1840–70

Date	Cultivated acreage	Number of recorded dairy cows	Dairy cows per 100 cultivated acres
c. 1840	24,186	2,241	9.3
1870	22,230	2,352	10.6

Sources: Phillips 1997.

wheat productivity between 1840 and 1870 still failed to approach the growth in food demand.

Although the available data do not furnish a full picture of change in Staffordshire agriculture, they point to general trends in the structure and productivity of its farming systems in the period 1840–70. In terms of output they demonstrate gains in both livestock and grain production; in terms of structure they reveal through the expansion of grassland a move to extend livestock enterprises. These developments may be seen as direct responses to the demands of the food market and as implementation of the strategies advocated by mid-century agricultural commentators. Yet the extent and rate of these improvements were of but modest proportions. They failed noticeably to approach output levels as dictated by the growth of population numbers and the rise in real wages; and the scale of restructuring was at a rate well below the upward movement in the price of livestock products. While agriculture in the county underwent a degree of change, it was insufficient to meet let alone satisfy the ever growing and varying demands of the Staffordshire food market.

This partial adaptation of agriculture in Staffordshire to market conditions may be seen as a product of a range of factors. First, capital outlay on the agricultural improvements that underlay the effective adoption of high-farming practices was limited. Of the improvements greatest stress in contemporary agricultural literature was placed on the need to underdrain land and to provide appropriate farm buildings. The emphasis on these two improvements stemmed from the fact that mid-nineteenth-century commentators regarded them as the bases of agricultural advance, the value of other inputs being largely dissipated without them. Nearly 60 per cent of the cultivated area of Staffordshire was subject to impeded drainage and would have benefited from underdraining. Removing excess soil water offered opportunities of raising yields of existing crops by about 10–15 per cent, of maintaining those yield advances, of allowing fertilisers to have greater effect, and of permitting greater flexibility in cropping. However, recent analyses have indicated that no more than 20 per cent of the cultivated land with impeded draining in Staffordshire was underdrained by 1870 (Phillips 1989; 1997).

Investment in farm buildings, in terms of both their number and layout, was central to attempts to increase agricultural output and to change the structure of farming systems. In particular, the keeping of greater numbers of cattle for either dairying or meat necessitated the provision of more accommodation specifically designed for livestock. Yet the amount of capital expenditure on farm buildings between 1840 and 1870 was of such an order as to restrict severely such developments. It has been estimated that outlay on the improvement in the period represented perhaps no more than 15 per cent of the sum required for the full renovation of the farm building stock in the county (Phillips 1997).

While the low level of provision of these improvements would have prescribed the extent of change in Staffordshire agriculture, explanation of their neglect is more difficult. Underdraining and farm buildings were expensive improvements involving sums of around £4 to £6 per acre. Because of their cost, by the middle decades of the nineteenth century landlords had become largely responsible for providing the capital for their adoption, with the tenant contribution being in the form of higher rent or payment of interest on the outlay. Despite their import-ance in facilitating change in agricultural output and farming systems, many land-lords in Staffordshire seem to have been unwilling to invest or doubtful of the benefit to be gained from investment in underdraining and farm buildings.

The imperative for large-scale investment may not have been readily appreci-ated by either landlords or their tenant farmers. The policies advocated by mid-century agricultural commentators to take fullest advantage of trends in the demand for and the price of food may have appeared too extreme. Indeed, the movements in the prices of agricultural products in the period may have been sufficient to render agriculture in the county prosperous without recourse to the costly adoption of new technical improvements and to radical changes in established farming systems. The measuring of levels of agricultural prosperity is difficult in the absence of extensive data on farm income in the county at mid-century. However rent per acre, long accepted as a reliable guide to the long-term movement in the overall prosperity of agriculture, may be used as a surrogate. Derived from assessments of agricultural land to Schedule B of the Income Tax, agricultural rent per acre in Staffordshire may be determined between 1842 and 1869 (Figure 6.2).

In 1842 rent per acre in the county was higher than the figure for England as a whole, suggesting an above average level of agricultural prosperity. By 1869 that prosperity had been enhanced, with rent per acre rising by 7 per cent from £1.74 to £1.87. Although this growth was slight, well below the rate that could have been achieved in the light of demand, nevertheless it was real and of an order to maintain the county's position relative to England as a whole. Because of price trends, landlords and tenant farmers could satisfy themselves with minor adjust-ments to farming systems and small productivity gains and not only maintain but improve their level of prosperity. Even with wheat, whose demand was limited, a

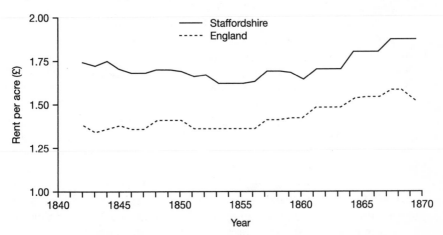

Figure 6.2 Agricultural rent per acre in Staffordshire, 1842–69
Source: Phillips 1997

profit could be made: despite the fall in its price and its sown area but because of yield increase, the value of its output in 1870 was greater than in 1840, even before taking into consideration savings arising from the cultivation of a smaller acreage. In such circumstances involving small-scale disruption of agricultural practices, pleas for thorough-going transformations of farming systems which promised even greater financial returns but which required higher outlays of investment in new agricultural technologies were likely to be unheeded.

If sufficient to ensure gains in prosperity, the improvements in grain and live-stock output did not reach levels to feed the growing numbers within the county. No available measure of agricultural output in Staffordshire between 1840 and 1870 approached the rate of population increase, let alone any other index of food demand. In its inability to feed its population, agriculture in Staffordshire was not unique: few counties in England were in a position by 1870 to be able to satisfy all food demands. Indeed, as Staffordshire was one of a handful of counties where currently available sources indicate a distinct increase in the overall area of grassland and green crops between 1840 and 1870, the move to livestock production may in other parts have been less developed (Phillips 1999). For Staffordshire, as for the country as a whole, the expanding shortfall between domestic agricultural production and home food demand had to be met by importation. Thus, while total meat consumption in the country increased by 19 per cent between 1841–50 and 1861–70, domestic meat production improved output by only 6 per cent, the gap in provision filled by overseas supplies. Again, imports of wheat rose from negligible amounts around 1840 to 46 per cent of the country's consumption in 1868–75 (Perren 1978; Smith 1968). The prosperity arising from agricultural developments in Staffordshire, as elsewhere in England, in the period 1840–70 should not obscure their severe limitations as an effective system of food supply.

Although relating specifically to Staffordshire, the findings from this study serve to illustrate many of the general problems of agricultural production and food supply in mid-nineteenth-century England and Wales. Consumers and producers can be seen to have differing requirements of the agricultural industry: the former sought a source of food, the latter a source of income. Although efforts were made to resolve these differences over the period, they were not eradicated. Landlords and tenant farmers were prepared to modify farming systems to an extent that they were able to sustain and augment their prosperity. However, the major reorganisations of farming systems and the large-scale investment in agricultural improvement that would have gone towards meeting more of the food demands of the consumer were never attempted. For agriculturalists such restructuring that did take place rendered the period a golden age; for consumers it provided an inadequate food supply system, making the use of overseas food supplies essential to sustain population growth. Indeed, closer reconciliation of the needs of producers and consumers of food in England and Wales was not achieved until a century later, after the Second World War, when the adoption of a policy of self-sufficiency, subsidies, and far-reaching technological advance in agriculture brought not only prosperity to farmers but also a sustainable supply of basic food requirements.

References

Caird, J. (1849) *High Farming under Liberal Covenants the Best Substitute for Protection*, 3rd edn.

Caird, J. (1852) *English Agriculture in 1850–52*.

Caird, J. (1880) *The Landed Interest and the Supply of Food*, 4th edn.

Collins, E.J.T. (1995) 'Did mid-Victorian agriculture fail? Output, productivity and technological change in nineteenth-century farming', *ReFresh*, 21, 1–4.

Ernle, Lord, (1961) *English Farming Past and Present*, 6th edn.

Evershed, H. (1869) 'The agriculture of Staffordshire', *Journal of the Royal Agricultural Society of England*, second series, 5, 263–317.

Ford, R.A. (1846) *How is the Farmer to Live?* Stone: privately published.

Grigg, D.B. (1989) *English Agriculture. An Historical Perspective*, Oxford: Blackwell.

Jones, E.L. (1968) *The Development of English Agriculture 1815–1873*, London: Macmillan.

Perren, R. (1978) *The Meat Trade in Britain 1840–1914*: London: Routledge and Kegan Paul.

Phillips, A.D.M. (1973) 'A study of farming practices and soil types in Staffordshire around 1840', *North Staffordshire Journal of Field Studies*, 13, 27–52.

Phillips, A.D.M. (1985) 'Settlement patterns in Staffordshire in the late eighteenth and nineteenth centuries', *Manchester Geographer*, new series, 6, 42–57.

Phillips, A.D.M. (1989) *The Underdraining of Farmland in England during the Nineteenth Century*.

Phillips, A.D.M. (1997) *The Staffordshire Reports of Andrew Thompson to the Inclosure Commissioners, 1858–68: Landlord Investment in Staffordshire Agriculture in the Mid-Nineteenth Century*.

Phillips, A.D.M. (1999) 'Revisiting a golden age: agricultural land use and cropping in Staffordshire, 1840–1870', in P. Morgan and A.D.M. Phillips, eds, *Staffordshire Histories: Essays in Honour of Michael Greenslade*, Staffordshire Record Society, 210–41.

Smith, W. (1968) *An Historical Introduction to the Economic Geography of Great Britain*, 2nd edn.

Sturgess, R.W. (1965) 'The response of agriculture to the price changes of the nineteenth century', Manchester University PhD thesis.

Thompson, F.M.L. (1968) 'The second agricultural revolution, 1815–1880', *Economic History Review*, second series, 21, 62–77.

Thompson, F.M.L. (1981) 'Free trade and the land' in G.E. Mingay, ed., *The Victorian Countryside*, vol. 1, 103–17.

Walton, J.R. (1990) 'Agriculture and rural society 1730–1914', in R.A. Dodgshon and R.A. Butlin, eds, *An Historical Geography in England and Wales*, 2nd edn, 323–50.

7 Envisaging the frontier
Land settlement and life chances in Upper Canada

Michael Redclift

Introduction

The previous chapter, by Anthony Phillips, examined one part of the British Isles – Staffordshire – where agricultural production made substantial advances during the nineteenth century. In the early part of the century most of the British population was rural, and much of it employed in agriculture. By the end of the century this had changed: most of the population was now urban. The centre of gravity thus shifted away from landowners, and towards industrialists and (eventually) industrial workers.

Food and diet were an essential part of survival – and sustainability – but the emphasis shifted away from production of food, towards its consumption. In terms of the sustainability of the land, as we have seen, the move was towards increased intensity of cultivation: there was little change in the land area cultivated, but dramatic increases in yields and, gradually, in the popular diet. The failure of the agricultural system to keep pace with the urban demands for food led, by mid-century, to the repeal of the Corn Laws, and the opening up of the frontier in North America, from which most basic grains reached the British market. The historical basis for sustainability thus moved from Europe to North America: the theme taken up in this chapter. For the migrants who went to Canada, and the United States, were part of the wider canvas in two ways – as the refugees from Europe's increasingly repressive rural societies, and as the forerunners of the people who contributed to the global bread-basket in the later nineteenth century, as the Great Plains were opened up in North America.

The impact of populations from the British Isles on the frontier of North America, is in some respects, a sequel to the narrative presented by Phillips. In many parts of the British Isles during the nineteenth century, and particularly in geographically marginal areas, poor rural people sought to achieve the fruits of economic development for themselves and their families, within the strait-jacket of established, and usually heirarchical, social institutions. In this context 'sustainability' refers not only to the capacity of natural resource systems to be renewed, but to the survival, and economic progress of the human populations that depend upon them.

The settlement of the Americas, in particular, suggests a different set of tensions. Here the European immigrant population was propelled by processes of 'expulsion' from the Old World, as well as attraction towards the New. Immigrants from Scotland, England and Ireland, in particular, settled in areas like Upper Canada (the region to the west of the Ottawa River) as the result of active encouragement by the British Government. They settled the frontier regions – at that time also known as 'Canada West' because it represented the most westerly point of settlement – and attempted to establish a basis in farming, lumbering and trade, in the face of considerable hardship. To many of them the wilderness they settled was 'hostile', it lacked the appearance of the 'civilisations' they had left behind. But if nature was hostile, it was also bounteous, and its conquest became a condition for the successful advance of rural communities and, in the course of time, an indigenous civil society. Many of the characteristics of Canadian frontier society in the 1840s and 1850s were reproduced in similar form in other settler societies, in parts of North, Central and South America, in the Antipodes, and in Southern Africa.

This chapter examines the different ways in which the frontier was envisaged by visitors and settlers to Upper Canada in the mid-nineteenth century, drawing on their own accounts of their experiences. It focuses in particular on the letters of Francis Codd, the son of a Norfolk parson, who arrived in 1846 in Pembroke, Ontario. The discussion of the social construction placed on the frontier forms the backcloth for the analysis of livelihood and sustainability with which the chapter ends. This analysis, in turn, enables a number of hypotheses to be generated about the relationship between the way in which nature is destroyed, and the way that different historical societies view the existence of natural 'limits', and sustainability.

Frontier expansion and the environment

In their short but incisive treatment of environmental history Beinart and Coates argue that: 'History, at its most fundamental level, records not only human interactions, but the incorporation of the natural world into the human world' (Beinart and Coates 1995: 9). This chapter examines this process in detail, by taking as an example that of Upper Canada in the mid-nineteenth century. The establishment of the frontier in predominantly 'settler societies' provides a useful illustration of the way that nature itself becomes 'socialised' through cultural processes, while social processes, including the establishment of civil society, are subjected to 'naturalisation'. Ultimately, it is suggested, the social construction placed on the physical frontier environment, particularly in the Americas, carries implications for the way in which modernist ideas of conservation and environmentalism have taken root at the global level at the end of the millennium. However, this argument, which forms part of a wider comparative project on frontier societies and sustainability, requires considerably more elaboration than can be provided in this chapter.

Upper Canada in the mid-nineteenth century

Upper Canada came into being after 1784, when about one-quarter of the Loyalist regiments that remained after the American Revolution moved northwards, into British territory. The numbers arriving were small: only about 7,500 Loyalists, and about 3,000 Free Blacks (Knowles 1992). On this modest base was constructed one of the largest immigrant populations of the early nineteenth century – by 1860 over one million new immigrants had arrived in Canada. Most of those who came to Canada, both from the United States and (more particularly) from Europe, were attracted by the promise of free land on the Ottawa River, where about 10,000 people had moved by 1812. Until that date most of the population of Upper Canada was still 'American', and made of considerable metal: 'The . . . Loyalists, who cut their way through the wilderness to the west of the province of Quebec, were true pioneers. Free land led them on into the Ottawa valley, and across the northern shore of Lake Ontario' (Andrea Koch-Kraft 1992: 151).

There were now two administrative divisions to Canada: east of the Ottawa River was Lower Canada (Quebec), while to the west lay Upper Canada. Between 1791 and 1851 the population of (British) Upper Canada rose from about 14,000 (mainly Loyalists) to almost one million people.

From the standpoint of the British Government in the early nineteenth century Upper Canada represented a bulwark against both the newly independent United States, to the south, and the French occupation of Lower Canada. John Graves Simcoe, the first Lieutenant Governor of the province, expressed the conventional Tory view when he stated that Upper Canada was 'a successful colony on the British model'. It was the objective of British policy to ensure that it remained so. In 1803 Thomas Talbot declared that Upper Canada remained a vital bastion, amidst worries about 'the growing tendency to insubordination and revolt' elsewhere in North America. It soon became official policy to encourage large-scale 'British' (to include Irish) immigration. Much of the immigrant population in the early nineteenth century came from North Connaught and North Leinster, two of Ireland's poorest, and most densely populated regions.

If the 'pull' of free passage, and free land, was attractive to British and Irish immigrants, developments in those countries provided a very considerable 'push'. One was the eradication of 'run-rig' agriculture, and another the Highland Clearances – many Highlanders made their way to the Maritime provinces in this period, and some travelled further, into Upper Canada. The most pressing factor, however, was the severe imbalance between the growing population on the land – in England and Ireland – and the limited possibilities of rural people acquiring land of their own. Those who arrived in the early nineteenth century were largely the 'detritus' of the industrial and agricultural revolutions; small farmers, weavers and small tradespeople, notably those who had survived the structural reorganisation of the Irish linen industry. Commenting on this immigrant force Knowles writes that, 'If immigration provided British North America with a much larger population, reasoned British officialdom, the colonies would be in less danger of being absorbed by their overpowering neighbour to the South' (Knowles 1992: 31).

These early colonists sometimes experienced spectacular success – the most notable example being that of Hamnett Kirker Pinhey, who secured a 1,000-acre land grant on the Ottawa River, and went on to become one of Canada's earliest political patrons.

The British Government offered free passage to Upper Canada for those from the 'troubled districts' of Ireland in the 1820s. The publicity for free passage helped to galvanise a large number of the independent migrants, and prompted Friendly Societies to sponsor them. Helen Cowan notes that, in this period, 'in Upper Canada the problem was how to fill its free grant townships with resident settlers, and at the same time develop its lumber trade' (Cowan 1961: 21).

The factor that turned the tide of British emigration policy, and served to encourage spontaneous migration, was the combination of poor agricultural conditions at home, culminating in the Irish potato famine, and the reports that increasingly emanated from the Land Companies that were established in this period. Upper Canada was increasingly seen as fit for those other than the destitute, such as Edward Gibbon Wakefield who, after spending three years in Newgate prison for abducting a young heiress, established himself in Upper Canada and promoted land sales to others from the 'old country'.

After 1830 the British Government made passage to Canada cheaper than to the American ports, and provided free transportation up the St. Lawrence River. The Whig Lord Durham spoke at this time of the need for a 'judicious system of colonization' which would swell the ranks of British migrants, and take the pressure off social conflicts at home.

Notwithstanding the limited success of assisted migration, those who journeyed to Canada in this period faced formidable difficulties, especially disease. In 1832 there was a cholera epidemic and, in 1847 alone, over 16 per cent of the 109,690 emigrants from Europe who left for Canada died before reaching Canadian soil (Koch-Kraft 1992: 152). Of those who arrived safely in Canada, a majority still headed south for the United States.

However, the policy of assisted passage did not prove as successful as anticipated by its British architects, who mistakenly saw colonial settlement as providing a solution to what were, in essence, British problems. Lord John Russel had, in December 1846, believed that emigration to Canada should be carried on only until the poor rates and the Emigration Commissioners (who were particularly eager for large-scale emigration) were satisfied that surplus population had been found a home. In 1848 he proposed a broad plan for local assistance, within the British Isles, for local emigration, but by this time it was already evident that a policy of *laissez-faire* was in operation.

The effect of land policy up to 1841 was not entirely satisfactory. Macdonald notes that: 'the land policy . . . permitted free grants in a most lavish and improvident manner, with the result that most of the Crown lands in the most accessible parts had been alienated, though [they were] far from being developed, or even occupied' (Macdonald 1966: 11). The goal was to avoid speculation in land while, at the same time, encouraging its 'development'; but by the mid-1840s most of the land that changed hands in Upper Canada was for speculative purposes.

The speculation in land was encouraged by the building of roads. Between 1845 and 1878 sixteen roads were opened up in the region to the west of the Ottawa River. The chief beneficiaries were the lumber industries associated with land development on the Madawaska River, one of the tributaries of the Ottawa. It was the revenue derived from lumbering that influenced the government's road-building programme, rather than the settlement of farming populations. Markets for farmers were geographically remote from the new farming areas, and these farms were less well endowed than those to the south. Land speculation provided the incentive for opening up a wilderness, although as in other speculative activities, the destruction of forest brought farmers in its wake.

Cowan notes that:

> The government-assisted settlements had proved expensive, the land grant system unique; neither method brought in the income needed to relieve the mother country of the cost of colonial administration. With these results in mind, and one eye always on the rival land sale system of the United States, the government began to appeal to man's purely selfish instincts, by making his reward depend solely upon his own efforts.
>
> (Cowan 1961: 113)

By the 1840s the forested areas of Upper Canada were becoming distinguished from the more prosperous farming areas to the south. Careless writes that: 'Canada West generally was so full of recent immigrants, and so much in the stage of extensive growth, rather than intensive growth, that its social structure was naturally ill-defined' (Careless 1967: 28).

On the Ottawa River a distinctive lumber community had developed, of whom some were 'farmer-lumberers', engaged on their own in logging during the winter, and farming during the summer. Many were 'shantymen', hired hands of Irish or French descent, an itinerant semi-proletarianised population which increased in the 1840s.

Careless provides a vivid portrait of this lumber community on the Ottawa valley:

> the . . . community showed the contrasts and inequalities of the fast-growing forest frontier. It was opulent and powerful on the one hand, crude, ramshackle and unruly on the other. Bytown, its capital, was the scene of frequent riots and head-breakings between rival Irish and Canadian lumber-men. In the early forties, in fact, a veritable 'Shiners' War' raged there, when the Irish element, known as shiners, sought to drive French Canadians out of the timber trade by force and intimidation, and met reprisals in turn . . . And yet this rowdy lumber town, later to be rechristened Ottawa, was blithely

urged by a local paper in February 1841, as a better choice than Kingston for
the capital of Canada.

(Careless 1967: 30–31)

Settlement and the market in Upper Canada

Before considering their attitudes towards nature and the environment, it is
important to establish the political economy of frontier life and, since it came to
play such a primary role in the subsequent expansion westwards, the role of the
market in early Upper Canada. Unlike the French who had settled in Quebec, the
frontierspeople west of the Ottawa River were not, essentially, subsistence producers. They were much more sensitive to the market than those to the east. Cole
Harris and Warkentin note:

> As a whole they were little interested in conservation, or the long-term man-
> agement of land, and sought to maximise short-term profits . . . settlement
> was a by-product of nineteenth century dislocation . . . where the human
> landscape had emerged from the wilderness . . . a great many gave the better
> part of their lives to clearing (the forest).'
>
> (Cole Harris and Warkentin 1991: 112–15)

Before large-scale settlement in Upper Canada in the 1840s the society – such
as it was – was Tory, highly stratified, and paternalistic, owing its allegiance to the
British Crown. What transpired served to change these structures and, ultimately
perhaps, these allegiances. In many ways the mid-nineteenth-century settlers were
behaving like classic liberals – seeking profits, chasing market opportunities, and
seeking to amass personal capital. Many of them became immersed in the business
of land speculation. They had not come to Canada to settle into self-sufficiency;
they had come to buy and sell. They hoped to sell their labour, or to farm pro-
ductively. To farm they needed land, and the land needed to be connected 'by a
passable road or by navigable water to the local village' (Cole and Warkentin
1991: 119).

Consequently, most early settlement in Canada west of the Ottawa River was
near the few roads that were built, where land was easier to clear, and surface
water provided. From its inception the settlement of land was tied to speculation
in it. If the soil became exhausted in the space of a decade the settlers were not
deterred: their objective was to amass the capital they needed to move on, to
prepare to colonise other frontiers. This endemic speculation in land, and the
difficulty of much of the terrain, meant that the settlement pattern was highly
irregular and dispersed, especially before the arrival of the railways in the 1860s.

The dispersed settlement, and the adventitious character of the population
established around it, meant that the attachment to land ownership quickly
assumed importance, even while land itself was changing hands. It is a mistake to
see the attachment to land, as a central core of 'property-owning democracy',
only in systems of mixed farming on long-established farms.

In a sense, dispersed settlement had been imposed on Upper Canada, and as a consequence the settlers who were highly critical of some aspects of government land policy, and the use of surveyors and land titles to establish freehold, were the stoutest defenders of the principle that made it 'virtually mandatory' in one commentator's view, to live on your own land (Cole Harris and Warkentin 1991). Land settlement was tied not only to ownership, but to land occupation.

If we find the beginnings of civil society in the clearing of land, and the establishment of title and occupation, we find it, too, in the values which this clearance gave rise to. Clearing forest and, just as importantly, preventing re-growth, produced

> an ingrained hostility to the forest and, eventually, a severe over-clearing of the land. English gentlemen visiting Upper Canada in the early and mid nineteenth centuries frequently contrasted their own sorrow with the settlers' exultation at the destruction of trees. For the one, trees were ingredients of a picturesque landscape; for the other [they were] a severe economic liability.
>
> (Cole Harris and Warkentin 1991: 135)

Forest was cleared to make way for a highly extensive system of agricultural production, and one which was adapted to the vulnerable market conditions that characterised much of Upper Canada. Land was cleared to cultivate wheat, and wheat was followed by fallow, the fallow period being followed by wheat. This wheat–fallow–wheat system was only replaced by more mixed farming as communications with urban centres improved, and local markets developed for more varied agricultural products.

Farming in the 1840s and 1850s in this area, was extensive, and market-orientated, in this respect acting as a precursor to the later Plains expansion. There was little emphasis on root crops or livestock production, although vegetable gardens and small-scale animal production were maintained for household use, in some cases. Any arguments about soil conservation techniques, or long-term, more sustainable planning of farming activities, or the conservation of the forest, when they were made at all 'fell on deaf ears' (Cole Harris and Warkentin 1991: 142).

The material practices of the newly populated areas reflected both market opportunities, and vulnerabilities, not least in the difficulties and obstacles that the natural environment presented. This was to find expression, at the time, and subsequently, in the way in which the frontier was envisaged: the discourses about nature and society that accompanied frontier life.

Discourses about nature

There are two discernible discourses about nature and the environment in the writings of visitors to Upper Canada in the mid-nineteenth century. The first is primarily concerned with the abundance of nature and the existence of plenty:

Nature is bountiful. The second discourse views nature as hostile to human beings, and asserts the need to 'tame' nature, to bring it within the fold of human activity and management. Visitors and settlers describe nature as wild and treacherous, far removed from the 'settled' landscape from which most British and Irish immigrants had come. The contrast is implicitly – and sometimes explicitly – with the British Isles.

These competing discourses provide contrasting accounts of early settlement, but they are also, in one sense, complementary. The implication is that by 'taming' wilderness, and bringing it within the compass of human institutions, one is able to enjoy the bounty, the rich resources, of nature. The environment which the colonists encountered – whether seen as abundant or threatening – was one which was linked to the human condition on the frontier.

Gerald Craig, in his account of *Early Travellers in the Canadas* (1955) identifies many travellers for whom the beauty of Upper Canada was a surprise. They commented on the vivid autumn colours of the forests, and on the magnificence of the trees they found. They also commented on the ferocity with which most settlers attacked the forest. An unknown writer of the period comments that 'In its present state in [Upper Canada] . . . should you wander a mile from a settlement, the face of man can hardly be viewed without an emotion of surprise, and every cultivated patch of ground is regarded as a trophy of his triumph over the desert' (Craig 1955: 38).

Issac Fidler was one of a number of Anglican clergymen who went to the United States in this period, disillusioned with the condition in his own country. Eventually he moved north to Upper Canada, in the 1830s, and commented on the potential wealth which the land afforded:

> So fertile is the soil of Canada, at its first cultivation after clearing, that an acre, upon which no more than one bushel is sown, will produce almost always between thirty and forty bushels. The first crop, with proper management, generally repays the purchase–money, the expense of clearing and fencing, the cost of seed, sowing and harrowing, and the expense of reaping, thrashing and carrying to the mill. In short, a prudent and industrious farmer may always calculate on being able to call the land he clears his own, by the first crop alone.
>
> (Fidler, 'The advantages of Upper Canada over the American States'
> in Craig (ed.) 1955: 91)

Another early account is by John Howison, who spent two years in Upper Canada between 1818 and 1820 (Howison, 'The Pioneer Society of Upper Canada After the War of 1812' in Craig (ed.) 1955). Howison records that the forest was cleared for '*improvement*', a term that had clear connotations for British visitors with the agricultural changes taking place in their own country. The occupation of the wilderness of Upper Canada was linked with the improvement of agricultural practices in England. Howison commented that in Canada there was, however, little attention to improved tillage as in England. The principal objective of farming was to establish the right to settle – which was dissembled and equated with *agricultural* improvement:

> The Canadians, in addition to their indolence, ignorance and want of ambition, are very bad farmers. They have no idea of the saving of labour that results from forcing land, by means of high cultivation, to yield the largest possible quantity of produce. Their object is, to have a great deal of land under *improvement,* as they call it: and, consequently, they go on cutting down the woods on their lots, and regularly transferring the crops to the soil last cleared, until they think they have sufficiently extended the bounds of their farms.
>
> (Howison, in Craig ed. 1955: 64)

Another traveller was C. R. Weld, who visited Upper Canada in the 1850s. He commented extensively on the scale of deforestation, and the rapid advance of lumbering. If the principal objective of the farmer was to establish his title, and to see the land improve in value, on the basis of high fertility; that of the lumbermen was to cut down the forest, in the knowledge that they would never be faced by a scarcity of timber resources:

> The question naturally arises, how long will the Canadian forests continue to meet the enormous demand for timber . . . A glance at the map of North America shows how small a portion of that vast country is included in this survey; so that although new channels of communication will be opened into the interior with the extension of commerce, it is not unreasonable to regard the supply of timber as almost inexhaustible.
>
> (Weld, in Craig 1955: 206)

The discourses about nature in the period are, not surprisingly, framed by the discourses of the home country, and the comparisons, either directly or indirectly, are usually with Britain and Ireland. In both these countries wild, uncultivated land was of poor quality, moor and mountain. Most of the cultivable land supported large populations of farmers – many of them very poor. The system of landholding was indistinguishable from the social structure of the society itself – with large, hereditary landholders at its head. For many of those who visited Upper Canada the most impressive sight was the apparently limitless extent of nature's bounty, which appeared to make few requirements of settlers, at least in the short term. For the settlers themselves, as we shall see, the establishment of their holdings, and the clearing of the forest, were necessary preliminaries to new livelihoods, made possible by their radically different life chances and, eventually, civil rights which they could only obliquely aspire to in England and Ireland.

Discourses on frontier society

We have seen how the appreciation of nature and the environment, among the people who colonised Upper Canada, was inextricably linked with their concern to establish their families and livelihoods in the face of natural hazards and official neglect. The letters and accounts written at this time maintain a tension between the promise of the unknown, and de-humanised, 'wilderness' and the 'savagery' of frontier society itself. What Anthony Trollope characterised as the 'roughness' of the frontier (in his book *North America*, published in 1862) was contrasted unfavourably with the 'civilised values' of Europe and the eastern seaboard.

Just as there were several accounts of frontier 'nature'; there were, similarly, several such accounts of frontier society. The first depicted the frontier of Upper Canada as one lacking in social refinement, coarse and 'uncivilised'. The other identified opportunities in the frontier that had not existed for the newly migrant population in their societies of origin. Upper Canada in the 1840s and 1850s demonstrated that in return for hard work people with modest means could survive and prosper.

These discourses on frontier society – like those on nature and the environment – took as their starting point a series of 'opposites': wilderness/settlement; violence/law and order; social refinement/uncouthness. However, in the texts we can also see evidence that the discourses on society and nature were assimilated into each other. The myth of 'civic virtues' which develops to 'civilise' nature itself also incorporates elements of the relationship *with* nature. This is expressed in terms such as the rough equality between individuals, the private ownership of land as a fundamental guarantee of civil rights, and the social mobility that came to characterise the frontier.

Craig comments on the way in which visitors to Upper Canada in this period emphasised 'sound educational standards, honesty in the government service and respect for good manners' as just as important as clearing the forest, or laying down railway tracks:

> Upper Canada . . . had everything in its favour. Here, in the heart of North America, the settler could enjoy the benefits of mild but orderly institutions, free of the high taxes, church tithes and harsh game laws that bore down so hard on the poor man at home. In Upper Canada there was an abundance of good land, available at nominal prices, and much more accessible than lands in far-off Illinois and Iowa.
>
> (Craig 1955: xxxix)

Similarly, Thomas Fowler, who journeyed through Upper Canada in the 1820s contrasted life in Britain, with its low wages, where wealth 'is drained in channels into the coffers of the great folks, as they are called, leaving the labouring classes and the agriculturist frequently without the necessaries of life', with British North America, where 'the farmers and the labouring class, all partake of, and have the wealth of the nation upon their hands' (Fowler 1955: 279).

Looked at from the perspective of those who might *lose social prestige by immigrating to Canada*, of course, the benefits of the frontier quickly become defects. John Bigsby in 1850 commented that 'the great defect in colonial life is the lower civilization which characterises it' (Bigsby 1955: 130). For this reason writers at the time sometimes cautioned prospective immigrants to Canada that although the province was ideal for the lower orders, whose life was a misery at home, emigration to Canada was of dubious value for those of easier circumstances. Bigsby cautioned that members of the 'better classes' who were 'content with plain comfortable mediocrity' were advised to come to Canada, a country which he described as 'a rude and rough place', which offered 'real temptations to wildness', somewhere where 'mental improvement is lost sight of' (Bigsby 1955: 131).

Over twenty years earlier John Howison had noted that, among recent settlers, there were 'absurd notions of independence and equality' (Howison 1955: 46). This was explained by another traveller, Professor Charles Daubeny in terms of the shortage of labour, which led 'to the tone of equality assumed by the lower class towards the higher' (Daubeny 1955: 192). A common reaction to the vulgarities of the frontier was that of Lieutenant Coke, who wrote that his English blood 'almost boiled in his veins' when he was placed at table with two servant women, a situation he might have anticipated in the United States, but had not expected in the British provinces (Coke 1955: 303).

These notable shifts in attitudes, and the removal of social deference from everyday life, were attributed not only to the absence of landlordism, and a rigid social hierarchy, but also to the reality of social mobility experiences by many of the migrants. John Howison commented that 'nine-tenths of the inhabitants were extremely poor when they commenced their labours, but a few years' toil and perseverance has placed them beyond the reach of want' (Howison 1955: 64). It was evident that the advantages to be derived from emigration to Upper Canada were not altogether 'chimerical' in Howison's view.

> No person, indeed, will pretend to say that the settlers whose condition I have described, are in a way to grow rich; but most of them now enjoy abundant means of subsistence, with the earnest of increasing comforts; and what state of things can be more alluring and desirable than this to the unhappy peasantry of Europe?'
>
> (Howison 1955: 61)

A similar sentiment was expressed by Issac Fidler, another visitor to Upper Canada in this period:

> Some of the advantages which emigrants of a lower order derive from change of country, is the comparative ease of mind which they possess. They are not tantalized by the presence of luxury from which they are excluded; and they find labour is a capital which yields them numerous and daily increasing comfort. They cannot obtain, nor can they reasonably look for, sudden

wealth. There is no region in the world, however fertile or well governed, that offers this to the generality of settlers . . . they see their flocks and herds increasing. They behold their families and houses supplied with more conveniences every day, and better furnished. They are not excluded, even at first, from the rights of citizenship, as in the States; nor from possessing real property, which immediately confers every political advantage, and which in most places can be cheaply purchased. . . . They find that their children are more easily provided for than in England, and will fill a higher place in the grades of society.

(Fidler 1955: 95)

These accounts of the way in which an apparent wilderness afforded ordinary immigrants with the possibility of both a livelihood and social advance are mirrored in the letters of one frontiersman, named Francis Codd, from whom we have detailed recollections, contained in the letters he sent home to his father, a vicar in Letheringsett, Norfolk. They provide a detailed source for this process of accommodation to frontier life, and the way that overcoming natural adversity became a first step towards the creation of a more or less indigenous civil society in Upper Canada.

Francis Codd and the frontier

In a series of letters written between 1847 and 1852 Francis Codd, the younger son of a Norfolk clergyman, describes the process through which Upper Canada was settled. He arrived in the town of Pembroke on 26 January 1847, from which he wrote about his journey from Bytown, wrapped in two buffalo skins, the only passenger on the sledge bringing the mail. He had found Bytown a flourishing town, expressing surprise that only twenty years earlier it had been a 'wilderness'. There were an estimated 8,000 inhabitants in 1847, and Bytown expected to be made a city during 'the next session of Parliament'. (Bytown became Ottawa, and eventually the capital of Canada, in 1851.)

He did not stay in Bytown for long; principally because there were already six doctors there! In the township of Pembroke the residents were keen to adopt him, and offered to send his travelling boxes on another sledge. He discovered people from Norfolk 'who spoke with a delightful Norfolk twang', who had arrived nine years earlier with 'nothing', and now owned the title to a 100-acre farm. His informant, a woman, observed that 'if the poor creatures at home knew what a place Canada is, it would be good for 'em'.

Francis is entranced by the beauty of the region, and by its isolation although: 'the country is not very civilised', he exclaims. After two weeks he comments on the cheapness of land: a half-cleared plot of 100 acres sold for £50. Two months later, having survived a hard winter he comments on the difficulties of survival in 'such an uncivilised place'. He loves the scenery and the lakes, but is owed money by most of his patients, who are lumbermen. The people are as wild as the country: 'wild when stroked; fierce when provoked. . . . I do not believe there is any part of the Western States so uncivilised as this. Certainly there is no part of Canada peopled by such savages . . . and no law, or civil power, within 100 miles to control them'. The barkeeper is threatened by the raftsmen, with his house being set on fire, and the lumbermen fight with *pillets* (large pegs for holding the timber together). Francis, the only doctor for miles around, tends the wounded.

A picture gradually emerges, however, of enormous energy and human potential. Francis notes that his father could settle on the land and, unlike an English parson with 'livings', he could be 'independent . . . The main art of living in Canada is to do with as little *cash* as possible, and if a man has a farm he can raise his own flour, pork, butter, cheese' he comments.

Several months later, in August 1847 Francis writes again from Pembroke. The lumbermen have been paid, and so his practice is generating some cash, at last. He has been out shooting partridges, and comes across a bear, at night, in the bush. He escapes, swims the river . . . and lives to tell the tale. By September he is telling his parents that although he wants to see them, he will probably 'leave old England for ever'. He picks blueberries and blackberries in pales, has his grog stolen by the Indians, eats wild plums and maple syrup. He still comments on the fights, especially those between Irishmen, which are particularly fierce. The potential of Upper Canada is such that 'if I had £200 and landed again in Canada, I would go into the bush and become a farmer . . . Do not think that I am tired of Canada because I grumble about Pembroke and call it an uncivilised hole, for there is as much difference between this part of Canada and any other part, as between Letheringsett and London.'

By January 1848 Francis is still in Pembroke 'and do not expect to get out of it till I get money . . . for the timber trade is worse than ever, and I cannot get paid in any shape as yet'. One of his patients commented to him that he was 'not half ruffian enough for this place'. Despite the uncivilised state of the woodsmen, Frances increasingly turns to the state of Canada in his letters during 1848, and comments that 'Canada is now changing from a colony to a powerful nation; her population already equals that of the white population of the States at the time of their rebellion.' There is a General Election, and Francis is sure that the 'Rads' will win. He comments that the previous Ministry were a 'set of sharks'. They had offered to sell the whole of Allumette island to a rich man, named Sparks, from Bytown, putting the poor squatters in jeopardy of being evicted. However, 'when Sparks came to see the place he found that he might as well buy a slice of the moon, for the squatters threatened to shoot him if he only went on the island to look at it'. He mentions a Mr Pinhey, who lives near Bytown who 'is now probably living in ten times the living and independence he did in England . . . he was lately made a member of the Legislative Council (the Canadian House of Lords) and has the title of Honourable for life . . . Had he stayed in England he would still have been a nobody, now he is the founder perhaps of a noble Canadian family, and owns the greater part of the township of March.' (This, according to the *Dictionary of Canadian Biography*, is Hamnett Kirkes Pinhey, landowner, politician and author, 1784–1857.)

During 1849 and 1850 Francis Codds' letters provide evidence that the wild and untamed country he appeared to disdain, was growing increasingly 'civil'. He comments on a Divisional Court being established in Renfrew, for collecting small debts, and his practice is 'slowly but steadily increasing'. He intends to vote Radical in the next election 'democratic principles of government are, in the present state of Canada, the best suited to her'. By May 1851 he is dissuading his younger brother Henry from migrating to New Zealand: 'what should he do in the most distant, and by all I can learn, most unpromising colony[?]'

He begins to see the merits of the civil society taking shape around him. He comments that: 'A magistrate in this country is, however, a very different animal from the same in England – he need not spend a dollar a year the more for being a magistrate – many of our magistrates are plain farmers who can just read and write decently, but their authority seems to be just as much respected as in England.' He advises his father that (if he were to come to Canada) 'I cannot see in what way you should be less comfortable in this country than in England, except as to servants . . . they are certainly a great plague here.'

The letters of Francis Codd provide a vivid picture of the frontier, both its society and its confrontation with 'nature'. There is evidence of what Bennett and Kohl (1995) call the strong 'booster spirit (of the frontier) . . . optimism about the conquest of nature . . . confidence in the technology of the late nineteenth century', a Promethean view of human potential. Writing about a later frontier generation Bennett and Kohl comment on the reasons Canadians have lauded their ancestors. The pioneers who 'fought the first battles with the environment [in Canada] and 'won' [were celebrated] because against all the odds, they succeeded in constructing viable human communities . . . in a kind of Siberia'.

Another point stands out clearly from both the detailed account of Frances Codd, and that of other early frontier people in Upper Canada. While some early settlers emphasised the need to bring civilised values to Upper Canada, they advocated doing so through settlement of the land, through the taming of wilderness. The paradox is that through exposure to wilderness their perspective on their own society changes. The environment – external nature – does not simply *reflect* social attitudes and values on the frontier, it comes to influence and transform these values. The discourses on frontier society and nature work dialectically, becoming discourses *between* society and nature, and in the process they serve to transform ideas about citizenship itself.

The interplay between these discourses, on nature and frontier society, has contributed significantly to the creation of both democratic forms and democratic *myths* in Canada. If 'sustainability' is used to refer not only to resource systems, but to the survival, and economic progress, of the human populations that depend upon them, then the way that the early Canadian frontier was *envisaged*, and the life chances of its human flotsam and jetsam, are closely linked. Indeed, the discourse between society and nature may have left its mark not only on mid-nineteenth-century Canada, through Canadian representations and myth, but arguably on the bigger picture we have inherited – on the global nature of environmental management today.

References

Bennett, J.W. and Kohl, S.B. (1995) *Settling the Canadian-American West 1890–1915*, University of Nebraska, Lincoln.

Beinart, W. and Coates, P. (1995) *Environment and History: The Taming of Nature in the USA and South Africa*, Routledge, London.

Bigsby, H. (1955) 'The shoe and the canoe, or pictures of travel in the Canadas', in Craig (ed.) *Early Travellers in the Canadas*. Macmillan, Toronto.

Careless, J.M.S. (1967) *The Union of the Canadas: 1841–1857*, Mckellard and Stewart, London.

Coke, E.T. (1955) 'A journey from Quebec to the Maritimes in 1832', in Craig (ed.) *Early Travellers in the Canadas*, Macmillan, Toronto, 303.

Cole Harris, R. and Warkentin, J. (1991) *Canada Before Confederation*, Carlton University Press, Ottawa.

Cowan, H.I. (1961) *British Emigration to British North America*, University of Toronto Press, Toronto.

Craig, G. (ed.) (1955) *Early Travellers in the Canadas*, Macmillan, Toronto.

Daubeny, C. (1955) 'Bytown and the Rideau anal in the 1830s', in Craig (ed.), 192.

Fidler, I. (1955) 'The advantages of Upper Canada over the American States', in Craig (ed.), 91.

Fowler, T. (1955) 'The Journal of a Tour Through British America', in Craig (ed.), 279.

Howison, J. (1955) 'The Pioneer Society of Upper Canada After The War of 1812', in Craig (ed.).

Koch-Kraft, A. (1992) 'Freedom in the North: Canada', in D. Hoerder and D. Knauf (eds) *Fame, Fortune and Sweet Liberty: The Great European Emigration*, Temmen, Bremen.

Knowles, V. (1992) *Strangers At Our Gates: Canadian Immigration and Immigration Policy 1540–1990*, Dundurn Press, Toronto.

Macdonald, N. (1966) *Canada, Immigration 1841–1903*, Aberdeen University Press, Aberdeen.

Weld, C.R. (1955) 'Lumbering and Farming near Peterborough in the 1850s', in Craig (ed.), 206.

Part 3

Geographical perspectives – the view from the South

8   Exploring dimensions of sustainability in Nigeria

A question of scale

Elsbeth Robson

Introduction

This chapter explores some dimensions of sustainability within Africa's most populous nation, Nigeria.[1] The discussion presented in the first part of the chapter reveals contradictions between apparent *unsustainability* with respect to Nigeria's ongoing economic, political and environmental crises; and *sustainability* at the sub-national level in regard to the distinctively unique cultural setting of Hausaland in North Nigeria. This major region of high population density appears to exhibit strong environmental, social and economic sustainability through highly productive intensive peasant farming, a distinctively cohesive social fabric and thriving agricultural markets. This is illustrated by empirical case study material from a Northern Nigerian village.[2]

Thus, this chapter illustrates the need to go below the macro-scale of the nation-state in discussing sustainability, in order to 'ground truth' the reality for particular regions. From the analysis it becomes clear that geography matters, scale makes a difference and variations in sustainable development within Nigeria are revealed. These reflect Nigeria's regional diversity of ecological and cultural zones from the southern humid tropical rain forests to the dry, dusty savannah of the North, occupied by Nigerians of over 100 language/ethnic groups, three major religious belief systems and stratified by class divisions from poor rural peasants to rich urban Mercedes-driving business managers.

The chapter examines evidence for sustainability in terms of livelihoods and life chances, proceeding further to look below the regional level to the scale of communities and households. Connections between socio-economic structures (especially gender), life chances and sustainable livelihoods are addressed. The empirical case study material from a North Nigerian Hausa village enables analysis of resource access, distribution and its implications primarily in terms of gender.

As households are cross-cut by numerous social and cultural divisions (between men and women, between young and old, married and unmarried) further contradictions in the regional picture of Hausa sustainability are found. Thus, life chances for women in Northern Nigeria (as elsewhere in the South) are constrained by their poorer access to goods and services necessary to ensure sustainable livelihoods.

Sustainability in Nigeria

There is an interesting contradiction in applying the concept of sustainability to, and within, Nigeria. In the late 1990s the Nigerian state undoubtedly has reached a highly unsustainable condition of political, economic and arguably also social and environmental crisis. Nigeria's seeming spiral into a deep abyss is a condition memorably described as 'the open sore of a continent' – title of Nigeria's Nobel laureate and brilliant commentator Wole Soyinka's (1996) public reflection on the Nigerian crisis from exile. Nigeria's collapse was acknowledged recently even by Nigeria's UN representative thus:

> After thirty-eight years of independence, during which we have had no more than ten years of civilian democratic rule, Nigeria's economic and social health is poor and there are indeed many threats to its very corporate existence. The multiple ailments afflicting the country include rising unemployment, collapsed educational and health systems, serious crime wave, continuous decline in the standards of living and overall quality of life; erratic public utilities and infrastructures, including water, transportation, electricity and telecommunications; serious brain drain; acute, prolonged and greatly embarrassing fuel scarcity and huge external debt burden.
>
> (Gambari 1999: 6)

Interestingly, while acknowledging many other aspects of crisis Gambari neglects to mention the environmental crisis of oil pollution and environmental degradation in the Niger delta. However, what is clear, even from the perspective of its international state representatives, is that in many respects at the macro-scale Nigeria lacks anything that could be described as a sustainable equilibrium. But in the midst of national collapse, endemic corruption, welfare deterioration, change, instability and uncertainty ordinary Nigerian villagers[3] continue to live their lives as military rulers come and go;[4] inflation escalates, debt multiplies and the Naira slides;[5] oil exploiting multi-national corporations (MNCs) pollute the waters of the Niger delta; the IMF tightens the screws;[6] the Commonwealth suspends Nigeria.[7] The list of calamities goes on, yet somehow daily lived experiences are sustained, life goes on, society is reproduced and somehow, the state, economy and environment continue to exist. Paul Theroux eloquently describes the daily process of sustainability from elsewhere in Africa, but it is equally a true picture of almost any Nigerian rural community:

> The village had an air of being industrious and yet nothing seemed to change. Was there something African in the way that all this energy and motion left no trace? Women carried firewood; big girls carried small children or else buckets of water; boys played or hoed the rows of corn; men squatted in groups, muttering and smoking pipes. Food was grown and cooked and eaten. Firewood was burnt. The buckets of water were emptied. The people

were sustained, and the achievement of the work was that life continued. All this effort was to hang on to life and remain the same.

(Paul Theroux 1994: 149)

As Theroux acutely observes at the micro scale of the village, at least in many parts of the Nigerian countryside crisis and unsustainability are not always so evident. Furthermore, many rural Nigerians, while agreeing that times are now hard and these days there is no money (*ba kudi*) readily admit that life chances and livelihoods have improved in living memory and certainly since British colonial rule (1903–60). This section of the chapter explores some of the contradictions between evidence for national and local sustainability (or its absence) in Nigeria. But first some context for the local level village study is needed:

Context

The original case study material analysed in this chapter originates from the distinctive ecological and cultural setting of Hausaland. This region covers a large portion of southern Niger and Northern Nigeria in the dryland savannah zone of West Africa. Precise delineations of Hausaland vary (Figure 8.1), but it is by any account a very large region covering around 310,000 km²,[8] with a population in excess of 50 million (Hill 1972; 1977: xii; Watts 1983: 25).

The Nigerian portion of Hausaland, more or less corresponds to Nigeria's drylands covering *c.* 260,000 km² with a population of around 34 million.[9] Across this area the Hausa language, dress, culture and ways of living predominate. Strictly, Hausa (and Hausaland) is a term based on linguistic, rather than ethnic definition because 'There are really no Hausa people; there are only people who speak the Hausa language and practise the Hausa way of life' (Fage 1955: 34). Hausa ways of life are closely bound up with local beliefs and practices of Islam. Islam first entered Hausaland in the late eighth century. The early nineteenth-century Sokoto Jihad powerfully campaigned for Islamic religious reform and left major impacts on today's Hausaland. Islamic praxis has significant implications for gender and impacts on men and women's life chances and access to livelihoods.

With respect to gender, the widespread practice of wife seclusion is one of the most important aspects of Northern Nigeria's cultural environment. The Hausa practice of *kulle*[10] involves the restriction of married women's movement outside the domestic compound. *Kulle* (i.e. seclusion) is not strictly expected of women who are divorced, widowed, or women beyond child-bearing years whether married, or not. The construction of a domestic domain whereby households via seclusion are organisationally separate from the 'public' world means that because women are confined and isolated within domestic spaces they depend on others (i.e. men) to represent and speak for them.

Empirical data on resource access collected during 1991–93 forms the basis of the rural case study analysed here. Zarewa, within Rogo local government, at the extreme south-west corner of Kano state (Figure 8.2) is the case study village.

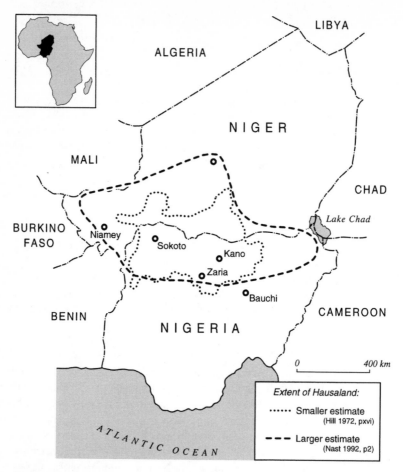

Figure 8.1 Hausaland
Source: After Nash 1992: 2

We begin the exploration of evidence of sustainability at national and local level by discussing the three dimensions of sustainability: environmental, social and economic sustainability. This conceptual framework has evolved from Munro's (1995) early attempts to elaborate the meaning of sustainability by identifying the complementary strands of ecological, social and economic sustainability.

Environmental sustainability

Environmental sustainability is ensured where the physical environment's contribution to human welfare and the economy can be sustained, that is, renewal of natural resource systems is maintained.

Since the 1970s Nigeria's development (sustainable, or otherwise) has

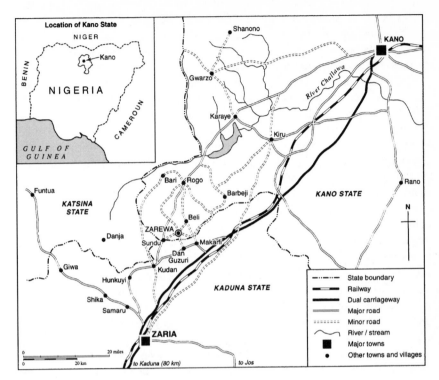

Figure 8.2 Location of Zarewa Village, Kano State, Northern Nigeria

depended on exploitation of one key environmental resource – oil, and will continue to do so for the foreseeable future. Other forms of direct production from the environment, namely farming, fishing and forestry stand a long way behind oil in revenue generation. But is the Nigerian environment being managed sustainably? Certainly the non-renewable oil deposits of the delta region continue to be exploited by MNCs (including Shell, Mobil, and Chevron), in compliance with the Nigerian state, in ways that create much environmental degradation and damage to human life chances and livelihoods through oil spills, pollution and gas flaring. Their actions deplete fish stocks, pollute the air, burn forests, make water undrinkable and land uncultivable while adversely affecting human health.

Arguably the problems of the delta region have stimulated grassroots environmental movements in Nigeria as highlighted by Movement for the Survival of the Ogoni People (MOSOP). There are also many other, less well-known, groups[11] actively protesting the actions of MNCs leading to a state of virtual low level warfare in the delta with military occupation, curfews, roadblocks and loss of life.[12] MOSOP gained international renown in 1995 with the arrest and eventual execution of its founder and leader Ken Saro-Wiwa. The Niger delta became a global focus for international environmental activists as Friends of the Earth launched a corporate campaign to pressure Royal Dutch Shell to 'clean up its oil production and end the company's complicity with the continuing human rights abuses of the Nigerian government' (FoE 1995). Elsewhere in Nigeria, beyond the delta there is no, or only nascent, environmental consciousness, but for most of the population so-called 'green concerns' are a luxury in the face of poor life chances and the daily pursuit of survival. The Nigerian Environment Study Team (NEST) is a small academically related NGO that has produced some reports and held conferences on the state of the Nigerian environment. A cynical perspective would suggest that such organisations are formed in response to the concerns of international agencies and the possibilities of funding from abroad. In other words, the environment becomes the latest bandwagon of the West onto which a handful of urban Nigerian academics struggling to survive years of inflation and dwindling salaries hitch themselves with hope of getting access to foreign exchange – a most scarce and sought-after resource.

There is evidence from elsewhere in Nigeria that environmental resources are being liquidated in the light of the ongoing economic crisis, for example, illegal logging of the small areas of remaining tropical hardwood forest, cutting down of urban and peri-urban trees for fuel by city populations deprived of kerosene for cooking. Thus, economic circumstances are forcing populations to short-termism in order to survive, even at the long-term cost of liquidating natural resources (e.g. chopping trees to make charcoal in face of need for cash (by charcoal makers) and urban demand for cheap fuel supplies as kerosene becomes either too expensive or unavailable).

Reducing the scale of focus to rural Hausaland, the environment sustains most of the population and their economic activity, as it does for most Nigerians. Northern Nigeria is a landscape of undulating plains with granite outcrops (*dutse*) on an underlying geology of Precambrian basement rocks with ferruginous tropical soils that are fertile and intensively cultivated (Mortimore 1970; 1968: no. 657, 300–03). Rainfall is highly variable and also uneven at a local scale. The study village of Zarewa lies approximately on the 1000 mm annual average isohyet (Mortimore 1989: 43).[13] Surface water in the region is limited to seasonal flow in most rivers, watercourses and marshy areas (*fadama*), but the water table is at varying depth and accessible in wells. Zarewa lies on the northern edge of the guinea savannah ecological zone[14] with natural savannah vegetation of grass, thorny scrub and denser woody vegetation along watercourses. The tropical savannah (Aw)[15] climate of the region has a single rainy season (*damina*) (Figure 8.1), which around Zarewa has a duration of approximately 145 days between May and September, the precise onset and cessation of the rains being highly variable. Agricultural activities centre on rain-fed, non-mechanised cultivation and closely follow the seasons (Figure 8.3[16]). Only *fadama* land is cultivated all year round being irrigated from wells by hand and occasionally with petrol pumps.

Nigerian Hausaland faces population growth and thus tension between meeting basic human needs for secure livelihoods and maintaining sustainable and productive agricultural systems. A fair degree of environmental sustainability has been possible because of the remarkable agricultural transformations in recent decades in Northern Nigeria (Smith *et al.* 1994; Adams and Mortimore 1997:

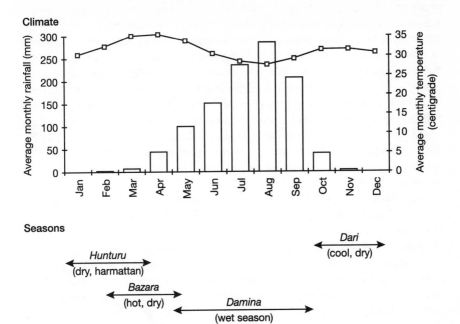

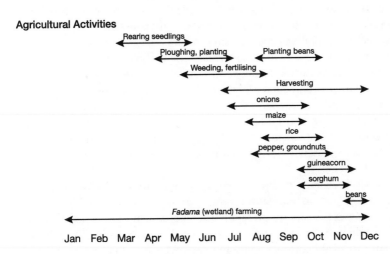

Figure 8.3 Climate, seasons and agricultural activities in Hausaland
Source: Akintola 1986 (figures for Samaru, Zaria)

no. 771). Goldman and Smith (1995) describe this as a 'green revolution', albeit little known and little recognised as such. At present Hausa peasant agriculture appears to be in a happy situation of environmental sustainability including good waste management[17] and a sustainable fuelwood situation.[18] Few women (12 per cent) and men (9 per cent) reported buying fuelwood in a household budget survey in Zarewa and expenditure on firewood was a very minor budget item. Most cooking in Zarewa depends on firewood collected from farm trees and shrubs, as well as grain stalks (*kara*) gathered from the fields at the end of the farming season. Farmers in Zarewa reported average ownership of 7.6 trees per household (6.2 on farmland, 1.1 in house compounds and 0.3 other). Zarewa resembles areas around Kano which are densely populated, but enjoy fuelwood self-sufficiency because farm trees have greater wood volumes and higher fuelwood yields than natural shrub and grassland (Cline-Cole *et al.* 1990: 64; 1995: 132). Whether the overall quality of environmental resources is being conserved, however, remains to be seen. Some possible areas for concern are examined below.

Areas of uncultivated 'bush' land (*daji*), which earlier this century were plentiful in Hausaland and provided grazing, hunting and gathering contributions to livelihood strategies, are shrinking as population growth leads to increasing demand for farmland. The trend towards almost universal permanent cultivation seems certain to lead to increasing conflicts and tensions between the sedentary farming Hausa and pastoralist nomadic Fulani, as well as permanent loss/transformation of environmental capital. The major loss in reduction of uncultivated land is the loss of gathering opportunities, in particularly traditional women's activities involved in securing medicinal products, forage foods in times of food shortages, fodder for domestic animals and firewood. Not only do the Fulani have less land on which to graze the cattle which are their livelihood, but the Hausa too face restrictions on where domestic sheep and goats may graze for the months of the year when fields contain growing crops.

The long-term impacts of chemical fertiliser use and reduction in, or eradication of, fallow periods on soil fertility and water quality are uncertain, but may well contribute to environmental problems in the next generation.

The impacts of climate change are being felt in this part of the West African guinea savannah bordering the Sahel. There is evidence that rainfall has been lower since the 1960s (Mortimore 1994: 379). Lower, patchier, less predictable rainfall is increasing the risks farmers face in maintaining agricultural livelihoods.

If the use of mechanical pumps for irrigation becomes more widespread with expansion of irrigated *fadama* farming the impacts on the water table could be disastrous for water supplies. At present households rely mostly on shallow hand-dug wells for domestic use. Larger nucleated rural settlements like Zarewa have government-provided boreholes and standpipes, which give intermittent service depending on the state of repair of the machinery, availability and presence of fuel, parts and personnel.

Examination at spatial scales below the level of the nation-state reveals regional variations in conditions of environmental management. While an eco-war is raging in the delta, Hausa peasants have sustainably intensified agricultural production to meet population growth.

Social (and ethical) sustainability

Social sustainability implies maintenance of social cohesion and reproduction of important social institutions and civil society. Ethical sustainability implies that human life (and other life forms) is valued in the present and protected for the future.

Within Nigeria there is plentiful evidence of social and ethical unsustainability. Threats to social cohesion and civil society are numerous, from ethnic and religious conflict, military rule, corruption, violence, abuse of human rights, Islamisation, and oppression of women, to name just a few. Certain key institutions are barely being sustained and face considerable struggle just to survive, especially trade unions, NGOs, education and health services, the media, legal, financial and political systems. In the face of uncertainty and crises some other institutions like households, churches[19] and other religious or ethnic associations are becoming stronger as means of securing livelihoods and improving life chances.

As one of the three major linguistic/ethnic/socio-cultural groups within Nigeria the (arguably hegemonic) Hausa-Fulani[20] exhibit strong social sustainability with deep roots long preceding the twentieth-century creation of Nigeria and subsequent decades of national turmoil. The Hausa emirs retain considerable authority, prestige and wealth tracing their rule back to pre-colonial Hausa kingdoms (the Hausa *bakwai*) originating from the formation of nascent states around the tenth to eleventh centuries. The hierarchical structure of traditional leaders extends to every village and ward including Zarewa. The ancient walled Hausa cities like Kano, Katsina, and Sokoto – huge urban centres today – can be traced back to urban developments as early as the fourteenth century. By the fifteenth to sixteenth centuries Islam was adopted and remains dominant today, although mixed syncretically with indigenous Hausa beliefs and practices. Villages like Zarewa[21] beyond the immediate fringe of the Hausa cities and in the south-west of Kano emirate were founded in the late eighteenth century or early nineteenth century by migrant traders from the north-east. With the exception of about half a dozen Igbo traders (and a visiting researcher) the village's 5,200 residents (as estimated by a household survey) are all Hausa-Fulani.[22] Village authority centres on the Sarki (chief), the Imam, the recently established court.

At least from the perspectives of outsiders (from nineteenth-century European explorers and British colonial servants, to more recent observers), therefore the dominant impression of Hausa society is one of changelessness (Shenton 1986: 120). Hausa culture and identity exhibited in language, dress, architecture, belief systems, ethnomedicine, music, agriculture, traditional crafts and the like remain powerful lived notions much written about by social scientists building on the classic early works of anthropologists M. G. Smith (1951, 1954 (reprinted 1981), 1955, 1960, 1962) and Polly Hill (1972, 1969, 1977).[23] Hausa society, however, even in villages like Zarewa is not entirely changeless and we should not read those anthropological texts as if it were. Although cultural markers such as Hausa dress, language (used in all spheres of public life including schools, TV, radio and newspapers), adherence to Islam, male circumcision of boys and facial scarification of newborn infants are more, or less, universal in the region, Hausa society is not reproduced unchanged from generation to generation.

Elderly Hausa residents in Zarewa tell of much change in their lifetimes and reflected in the stories told them by grandparents.[24] Hausaland has undergone lots of changes over time, but with strong parallel threads of continuity. The Hausa political economy continues to rest on agriculture and trade. The details change – the fortunes of cotton as a cash crop wax and wane, millets and sorghums have been overtaken by maize as staple grain, groundnuts come and go, bicycle spares are more in evidence than donkey panniers, factory-printed cloth has largely replaced hand-woven textiles. Other traditional handicrafts of weaving, indigo dying and calabash carving, which thrived earlier this century in Zarewa, have all but died with the introduction and widespread trade in manufactured commodities.

Hausa society is extremely resilient and stable, having survived and evolved over many centuries and will doubtless continue to do so. Certain key social institutions, especially Islamic ones, are being strengthened within Hausaland at the present time in the current climate of growing Islamism (part of a global trend). However, Hausa society is not immune to disruptions to peace and stability characteristic of Nigeria's present unsustainable condition. So-called religious, or Christian–Muslim, riots have occurred during the last decade or so in the northern region. These incidents can be read diversely as resulting from growing fundamentalism, as belonging to the volatile ethnic relations that currently characterise and threaten Nigeria's federal character where fights over scarce resources erupt, and/or as deliberate state sponsored violence to justify continuance of military rule.

At village level, society continues to be reproduced. The annual cycles of Muslim festivals (Eid-el-fitr, Eid-el-kabir (*sallah*), the birth of the prophet (*mauludi*), Ramadan) are observed. Marriages occur in plenty after harvest, deaths are mourned, births celebrated with naming ceremonies (*sunna*) and divorces marked by women's ritual separation (*iddah*).

However, many state-provided facilities in villages like Zarewa reflect the sustainability crises of the state. During fieldwork Zarewa's local electricity supply had not worked in years, its government primary schools were barely functioning with buildings in chronic state of disrepair, unfilled teaching posts, low enrolments, and poor pupil achievements. The dispensary was without drugs, the rural bank was not flourishing (it has since closed) and potholes in the main village access road were increasing with each rainstorm.

But other routes of patronage meant some village institutions were better supported. The recently established Islamiyya primary school (taking predominantly girls and with an explicitly religious curriculum) was growing and thriving. It had new school buildings built as a spin-off from the prestigious Challawa Gorge Dam project elsewhere in the local government area. A new clinic was opened in 1993; gateways at the village entrances and a market-place followed shortly afterwards. In gaining these facilities Zarewa owes much to strong representation of villagers among the local government officials.

A condition of ethical sustainability is that human beings are valued today. Discussion of ethical sustainability, as far as people are concerned, takes us into the realm of human rights. Nigeria has a poor record in this respect. The highest profile cases in recent years have been the detained chief Moshood Abiola (widely believed to have won the presidential elections of 1993 and who eventually died in detention in 1998), the murder of his wife Kujirat (thought by many to have been assassinated by the state) and the Ogoni hangings. To take just an example, Nigeria has long been known for its flourishing, diverse and outspoken print media, but in recent years journalists have been harassed, detained and worse; newspapers have been banned, magazine offices burnt down and so on. Ethical sustainability does not reign in society at large where corruption, crime and violence are endemic and often perpetuated by the state. However, the situation at the end of the 1990s is looking slightly more positive under Abdulsalami Alhaji Abubakar than under Sani Abacha. Since taking office the current president has released some detainees.

At village level in a rural community like Zarewa evidence of the unethical nature of the state is less obvious. Crime is low, but more frequent than in the past. Older villagers recall how bags of grain or cotton could safely be left outside a house, or chilli peppers left drying on a roof, but now come nightfall valuable agricultural commodities must be locked away inside. A vigilante group of the village young men (armed with menacing clubs and whips) patrol the village each night. In recent years Zarewa has acquired a court. Zarewa, however, is not immune from the national culture of backhanders whereby almost every provision of goods or services by the state at national, state or local government level is tendered by giving out contracts to individuals who then subcontract to other individuals and so on, each person creaming their own share. Around Zarewa the local government gives such contracts occasionally for things like culvert construction and they are generally snatched up by young male élites (e.g. educated sons of chiefs/ward heads, large traders, local government officers). Lack of gender equality and low status of women is a major aspect of the weak ethical sustainability in Nigeria in general and is well illustrated with the case material from Zarewa specifically.

Gender inequality in Zarewa reflects unequal life chances. While life chances for Nigeria as a whole are poor, they are worse in the Northern regions and the rural areas. Thus, Nigeria's infant mortality rate has improved from 99 per 1,000 live births in 1980 to 78 per 1,000 live births in 1996 (World Bank 1999: 203). Infant mortality figures for the Hausa region are much higher: in the rural town of Malumfashi 137.2 per 1,000, despite gross under-reporting of infant deaths (Verma and Singha 1982: 193). In the villages of Matsai and Bari infant mortality was estimated to be as high as 200–50 per 1,000 live births, with a ratio of dead to living children of 0.55 and an average of about four surviving children per woman (Lockwood 1994). For Zarewa the ratio of women's dead to living children is 0.53, i.e., only a two-thirds survival rate, which resembles other studies in the region. But while life chances are poor in the early years the risks are more or less equal for girls and boys and no demographic sex bias is apparent. Gender inequality becomes evident in childhood as children begin to be treated socially as little women or little men, rather than merely as infants. For example, discrimination is very evident with respect to schooling.

Provision of education in Zarewa is gendered. The many informal Koranic schools (*makarantar allo*), the two government primary schools and the Islamiyya primary school are open to boys and girls, but the junior secondary school is for boys only. Education can contribute to sustainable development and equal education for boys and girls is a significant ethical component of sustainability. At all scales of analysis, for Nigeria, the Northern region and Zarewa village, the education status of women is lower than that of men. Nigeria's national adult literacy rate (1995 figures) is 47 per cent for women (but 67 per cent for men) (World Bank 1999: 193). Across Kano State 24 per cent men, but only 18 per cent women, are literate (Nigeria 1995: 95). In 1987 in Kano State girls accounted for only 32 per cent of total primary school enrolment; just 8 per cent of primary school teachers and 8 per cent secondary school teachers in the state were women, while girls formed only 16 per cent of secondary school pupils and women were a mere 7.4 per cent of students at Bayero University Kano (Kano State 1987; Nigeria 1995: no. 734, 90).

In Zarewa, with the odd exception of a few married women who grew up outside the area, all adult women are illiterate. Even if they attended some schooling in childhood, reading and writing is not part of women's everyday lives. Girls constitute just 25 per cent of the pupil enrolment at Zarewa's primary schools (1992 figures). While girls are a small minority in the mainstream 'Western' education schools in the village, they outnumber boys attending the village's Islamiyya Primary School (Figure 8.4). A large portion of children in Zarewa do not attend primary school, only Koranic school. The continuing Nigerian economic crisis of the 1980s and 1990s (Olukoshi 1993; Mustapha 1991) (Figure 8.5) has meant fluctuating enrolments (Figure 8.4[25]) and deterioration in the quality of education. Furthermore, the degree of functional literacy pupils attain in any of the schools is questionable because of the poor quality of teaching resources, crowded classroom environments and the inclusion of three languages in the curriculum (Hausa, Arabic and English).

In 1992 the two Zarewa primary schools had nearly three times the number of boys than girls (Figure 8.4).[26] In 1991 forty-three boys leaving the primary school continued their education at Zarewa Junior Secondary School and six boys went to secondary school elsewhere. Only two girls (daughters of a school inspector and a head teacher) passing out in 1991 got places at boarding secondary schools elsewhere in Kano State.

In contrast to the mainstream primary schools, Islamiyya schools have a very high proportion of female enrolment. In 1992 a total of 268 pupils were attending the Islamiyya School – nearly four times as many girls (176) as boys (43) (Figure 8.4).[27] Islamiyya primary schools provide only four years of primary education, instead of six years, and do not prepare pupils for admission to secondary school. Thus, they do not conflict with the marriage and subsequent seclusion of

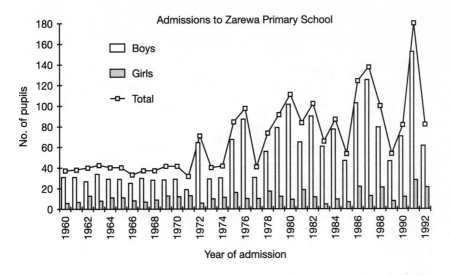

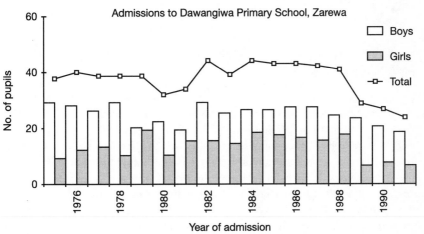

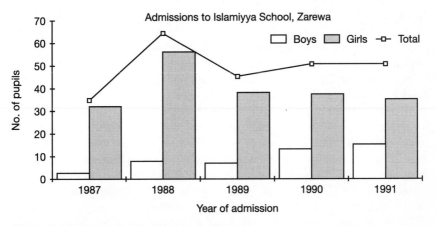

Figure 8.4 Primary school admissions in Zarewa

Figure 8.5 The condition of Nigerian education
Source: Daily Times (Nigeria) 13 May 1993

girls shortly after puberty, thereby providing schooling considered to be in tune with local customs and beliefs.

Often parents send their daughters to an Islamiyya primary school, while sending their sons to the mainstream primary schools (although many children do not attend school at all). A reason for the evolution of Islamiyya schools and parental preference for this type of schooling for daughters would appear to be that the education provided by Islamiyya schools is not seen as threatening, unlike 'Western' education that is perceived as a threat to early marriage, *kulle* and pre-marital chastity for girls (Stock 1985: 283). An Islamiyya school education is perceived as preparing girls to fulfil the norms of what 'good' women should be and thus, perpetuates prevailing gender ideologies. For example seclusion, as part of gender praxis, is maintained in part by women's conformity, so its continuation as a social institution depends on the socialisation of young women (and men) to act to ensure its reproduction. Islamiyya schools are one of the new ways of sustaining gender ideologies, including seclusion and its praxis.

All boys and girls in Zarewa attend Koranic school (*makarantar allo*) from an early age, but generally boys study the Koran for more years than girls. In the village only men complete recital of the Koran from memory. There is a regional dimension to this indigenous form of education, which has been in existence for centuries. *Almajiri* (Koranic students) come to Zarewa to study under a Malam, especially during the dry season (some may stay in the village for some years); likewise, many boys from Zarewa are sent by their parents to study the Koran elsewhere and may not return for five years or more. The high mobility levels required in acquiring religious knowledge is a barrier to women's participation beyond a basic level. At the age boys go off to study girls are approaching marriage, if not already married and secluded.

So overall access to formal education is highly gendered in Northern Nigeria. Rural areas get lower enrolments and poorer quality education than urban areas and men get more access to education than women. Patterns in unethical and unsustainable development in access to education are cross-cut by variations of scale, region and gender. Girls' poor access to education is followed by other related aspects of gender inequalities which for reasons of space can not be explored in detail here but include lower marriage age than men, high maternal mortality and less access to health care. In turn women's poorer life chances than men lead to poorer livelihood opportunities.

Hausa women and men largely pursue separate and individual livelihood strategies and do not contribute to a common purse, but share complex responsibility arrangements for household provisioning. Women generally have lower income earning capacities in trading occupations than men because of exclusion from long-distance trade, time spent child bearing and limited access to market-place trade. Women are excluded from farming due to limited access to and ownership of land, gender divisions of labour which exclude women (except occasional girls and widows) from on-farm field labour. All these factors which reduce women's opportunities to secure their own livelihoods are more or less due to the socio-cultural constructions of gender and praxis of *kulle*. Overall Hausa households successfully maintain livelihoods even in the face of external economic pressures, although there are occasional, but recurrent, crises of survival related to food shortages and famine (e.g. work of Watts, etc.).

Therefore this case study shows that as in various societies (particularly in the South) women's life chances are frequently constrained, not least through access to goods and services necessary for sustainable livelihoods. In Hausa rural Northern Nigeria many women are denied access to public spaces thus limiting their opportunities to earn income and secure livelihoods for themselves and their children. Men enjoy greater livelihood opportunities through being enabled by patriarchal ideology and praxis to dominate public space farming and trading spheres and activities.

As a rural community, Zarewa, like much of rural Hausaland is a tight-knit, highly structured, cohesive society with stability that has been sustained for centuries. Zarewa villagers will continue to practise the Muslim-Hausa way of life, integrating some outside values (e.g. communication in English, use of western scientific medicine) and disregarding others (e.g. gender equality), as it has in the past. Hausa society remains highly patriarchal, patrilocal and patrilineal, being strengthened in this regard by the climate of increasing Islamism reinforcing the practice of *kulle*, for example. So while Nigeria fares badly at the state level in terms of social sustainability, indigenous aspects of Hausa communities do better, but at neither scale is there much evidence for ethical sustainability.

Economic sustainability

Nigeria is clearly not enjoying economic sustainability. Its economy exhibits highly unsustainable development in view of its falling exchange rates, escalating national debt, low GDP, hyperinflation, shocks and crises in recent years (Table 8.1). Economic crisis from the mid-1980s led to a raft of structural adjustment policy measures first adopted in 1986 by President Babangida.

Altogether the Nigerian economy presents a picture of decline and crisis within which in recent years life chances and secure livelihood opportunities have declined for Nigerians. The informal sector, especially trade (including cross-border smuggling and the sale of petrol bypassing filling stations), is thriving; as is crime and corruption.

Hausaland belongs to the larger state of Nigeria, but it is helpful to move away from the macro-statistics to consider whether the Hausa people are surviving and progressing economically. The economy of Hausaland is based on agricultural production and trade in agricultural commodities, as it has been for many centuries. Both farming and trading as areas of economic activity continue to thrive. Agricultural producers, it can be argued, have been among the few to benefit from SAPs as farm gate prices have increased, although the benefits are partially offset by increased prices for farm inputs (especially

Table 8.1 Nigerian economic statistics 1985–98

Year	Exchange rates[a] (₦ per US$)	Consumer prices (Index: 1990 = 100)	Interest rates (Lending rates[b])	Foreign debt (₦ = millions)	GDP (US$ per capita)
1985	0.89	34.1	9.43	17,290	850
1986	1.76	36.0	9.96	41,452	423
1987	4.02	40.1	13.96	100,787	267
1988	4.54	61.9	16.62	133,956	305
1989	7.36	93.1	20.44	240,033	327
1990	8.04	100	25.30	298,614	337
1991	9.91	113.4	20.04	328,051	330
1992	17.03	163.4	24.76	544,264	316
1993	22.07	256.8	31.65	633,144	302
1994	22.00	403.3	20.48	648,813	383
1995	21.90	696.9	20.23	716,866	—
1996	21.88	901.0	20.32	617,320	—
1997	21.89	977.9	20.41	—	—
1998	21.89	1037.9	—	—	—

Sources: IMF (1997 (Sept.), 534–37, 1992, 540–43, 1998 (Dec.), 550–53, 1998, 682–85).

Notes
The data show an economic decline which progresses steadily from year to year; however, the averaged annual data smoothes out the extreme fluctuations in prices and exchange values which continue to occur on a daily, weekly and monthly basis, making the economic climate in which individuals and households struggle for sustainable livelihoods even more insecure.
[a] Central Bank official exchange rate.
[b] Lending rate: rate charged by banks on first-class advances.

fertiliser) and other retail commodities. Around Zarewa there are huge bustling market-places which act as bulking points for grains, vegetables, sugar cane and chillies grown by Hausa farmers. Long-distance trade in particular is very profitable. However, within this picture of economic sustainability some are benefiting and thriving more than others, for example small-scale traders make smaller profits than large-scale and long-distance traders, male traders dominate the most financially rewarding long-distance trade, while women are restricted to intra-village trade.

Hausa peasants are hard working, resilient, resourceful and quick to respond to economic opportunity. While not immune to the macro-scale crises in the Nigerian economy (e.g. inflation and fuel shortages affect almost everyone) farming and trading are sustaining rural Hausa populations.

Conclusions

The chapter has briefly examined aspects of sustainability in Northern Nigeria in part through analysing the ability of rural populations to secure and maintain livelihoods by using and transforming resources. As an examination of the current rural Hausa 'way of life', it is important to recognise this is not something static but what is presented here is a limited snapshot because the data available are not all encompassing and were collected at a particular place and point in time. In other words, there is no single Hausa way of life, only modes of livelihood or livelihood strategies which vary and change across time (historically) and space (regionally). Thus there remain today and have been differing urban and rural ways of maintaining Hausa livelihoods for centuries, while during this century in particular both rural and urban livelihoods have been transformed. Furthermore, at particular points in space and time livelihood strategies differ markedly across class, and within households are further differentiated by age/marital status/ stage in the life cycle and gender. Not all these variations could be explored in detail here.

The environmental transformation of Hausaland has in many ways been a triumph of human mastery over nature, contributing to the common good while socially sustainable, providing subsistence and increasing standards of living for growing numbers of people over time. The case study explores and presents a picture of lived experiences of sustainability – in other words locates sustainability in this particular regional context in the everyday life of Northern Nigerian villagers. In many ways the rural Hausa way of life, that is livelihoods, exhibits both environmental and cultural integrity. Its sustainability lies in defence of existing values, for example good environmental practice (such as planting and protection of economically valuable farm trees, use of domestic animal manure as organic fertiliser, very limited low-technology irrigation using little groundwater, etc.). But the Hausa savannah environment of rural Hausaland remains one of delicate natural resource balance that may yet be threatened by complex interlinked physical, demographic and external economic forces, most notably climate change, population increase, economic crisis and SAPs, as well as current agricultural practices, especially reliance on chemical fertilisers to maintain agricultural yields.

Notes

1 Nigeria's total population in 1997 was estimated to be 118 million, more than one-fifth of the population of sub-Saharan Africa (World Bank 1999: no. 762, 191).
2 Funding for this project is acknowledged from the Economic and Social Research Council, the Dudley Stamp Memorial Fund and the University of Oxford, namely, Linacre College, the Beit Fund, the School of Geography, and the Graduate Studies Committee.
3 The majority (59 per cent) of Nigeria's population lives in rural areas (World Bank 1999: no. 762, 193).
4 In the last few years Nigeria has had three different military rulers: Ibrahim Babangida (1985–93), Sani Abacha (1993–98) and Abdulsalami Alhaji Abubakar (since 1998).
5 See Table 8.1.
6 IMF and World Bank imposed structural adjustment policies have been implemented since July 1986. For elaboration see Olukoshi 1993: no. 745.
7 Following the execution of writer Ken Saro-Wiwa and other Ogoni activists on 10 November 1995 despite intense international pressure in the face of flagrant human rights abuses.
8 Even if only half of Hausaland is reckoned to be in Nigeria (and estimates vary considerably), the region covers nearly one-quarter of the land mass within Nigeria's state boundaries.
9 Population estimate for 1999 based on 1991 census figure of 28 million (Cline-Cole 1995: no. 613, 174) and a population growth rate for Nigeria of 2.9 per cent 1990–97 (World Bank 1999: no. 762, 195).
10 *Kulle* literally means to lock up and refers to the locking up of wives in their husbands' houses. '*Kofa tana kulle*' – the door is locked. '*Tana kulle*' – she is in purdah (Abraham 1949: no. 665, *Dictionary of the Hausa Language*, 555).
11 For example Ijaw Youths Council (IJC), Movement for the Survival of the Ijaw Ethnic Nationality in the Niger Delta (MOSIEND), Niger Delta Volunteer Force.
12 Details of the unfolding Ijaw crisis are documented by Niger Delta Human and Environment Rescue Organisation (ND-HERO 1999: no. 770).
13 Although rainfall at Zarewa is less than that recorded 50 km further south at Zaria (annual average 1062 mm), it is equally variable – the long-term average annual minimum at Zaria is 287 mm and maximum 2,172 mm (Akintola 1986: no. 654).
14 For a detailed description see Jackson (1970: no. 744).
15 According to the Köppen-based classification scheme of world climatic zones.
16 Source of rainfall data: Akintola (1986: no. 654).
17 Waste disposal is relatively good in rural areas where most organic waste matter is used as fertiliser and most inorganic waste is burnt. As in most low-income communities there is little waste as many non-organic materials are reused. However, in common with other African countries non-biodegradable plastic bags used widely by retail traders are the notable exception and discarded 'plastics' create unsightly litter along streets, roadsides and open land where they also pose a hazard to small children and grazing animals.
18 Detailed empirical data successfully demonstrates the absence of a fuelwood crisis in the Kano region (Cline-Cole 1990: no. 773).

19 Especially pentecostal ministries (Marshall-Fratani 1998: no. 774, 283).
20 Explain Hausa-Fulani.
21 Zarewa lies 125 km south-west from the state capital and largest city in Northern Nigeria, Kano, and 50 km north-north-east from the nearest urban centre of Zaria.
22 In the 1963 census 94 per cent of the population in Kano state were recorded as Hausa-Fulani, a situation which has not changed for rural areas although urban centres have become more ethnically diverse (Kano State 1987: no. 610, 84).
23 Some central characteristics of the Hausa way of life are summarised in *Hausa Customs*, much reprinted since its first publication (Madauci 1968: no. 595).
24 Ten elderly women and twelve elderly men in Zarewa were interviewed during fieldwork about their life stories.
25 Data provided by school headteachers in Zarewa.
26 The reluctance to send girls to primary school is not new or confined to Zarewa. In 1976–77 85.6 per cent primary school students in Hadejia town were male (Stock 1985: 280–81).
27 Higher female than male enrolment in Islamiyya schools is reported as a general finding across Hausaland (Bray 1981: 63).

References

Abraham, M. R. C. and Kano, M. M. (1949) *Dictionary of the Hausa Language*, London: Crown Agents for the Colonies.

Adams, W. M. and Mortimore, M. J. (1997) 'Agricultural intensification and flexibility in the Nigerian Sahel', *Geographical Journal*, **163**, 2: 150–160.

Akintola, J. O. (1986) *Rainfall Distribution in Nigeria 1892–1983*, Ibadan, Nigeria: Impact.

Bray, K. (1981) *Universal Primary Education in Nigeria: A Study of Kano State*, London: Routledge & Kegan Paul.

Cline-Cole, R. A. (1995) 'Livelihood, sustainable development and indigenous forestry in dryland Nigeria', in *People and Environment in Africa* (ed. Binns, T.) London: John Wiley, pp. 171–185.

Cline-Cole, R. A., Falola, J. A., Main, H. A. C., Mortimore, M. J., Nichol, J. E. and O'Reilley, F. D. (1990) *Wood Fuel in Kano*, Tokyo: United Nations University Press.

Fage, J. D. (1955) *An Introduction to the History of West Africa*, Cambridge: Cambridge University Press.

Friends of the Earth (1995) *Corporate Campaign: Royal Dutch/Shell in Nigeria* (accessed 03/02/99) www site: http://www.foe.org/orgs/ga/shell.html.

Gambari, I. A. (1999) 'Nigerian foreign policy in a changing domestic and external environment', Presentation at Day Conference on Political and Economic Change in Nigeria, 26th January 1999, Oxford University.

Goldman, A. and Smith, S. (1995) 'Agricultural transformations in India and Northern Nigeria: Exploring the nature of green revolutions', *World Development*, **23**, 2: 243–263.

Hill, P. (1969) 'Hidden trade in Hausaland', *Man*, **4 N.S.**, 392–409.

Hill, P. (1972) *Rural Hausa: A Village and A Setting*, London: Cambridge University Press.

Hill, P. (1977) *Population, Prosperity and Poverty: Rural Kano 1900 and 1970*, London: Cambridge University Press.

International Monetary Fund (1992) *International Financial Statistics Yearbook*, Washington: International Monetary Fund.

International Monetary Fund (1997 (Sept.)) *International Economic Statistics*, Washington: International Monetary Fund.

International Monetary Fund (1998 (Dec.)) *International Financial Statistics*, Washington: International Monetary Fund.

International Monetary Fund (1998) *International Financial Statistics*, Washington: International Monetary Fund.

Jackson, G. (1970) 'Vegetation around the City and nearby villages of Zaria', in *Zaria and Its Region* (ed., Mortimore, M. J.) Occasional Paper no. 4: Department of Geography, Ahmadu Bello University, Zaria.

Kano State (1987) *Statistical Yearbook 1987*, Kano, Nigeria: Kano State Ministry of Finance and Economic Planning.

Lockwood, M. (1994a) 'Rural Household Surveys', 9th March 1994, Seminar at Centre for the Study of African Economies, University of Oxford, Oxford.

Lockwood, M. (1994b) Households and Risk in Northern Nigeria, Research Seminar (31/5/94), School of Geography, University of Oxford, Oxford.

Madauci, I., Isa, Y. and Daura, B. (1968) *Hausa Customs*, Zaria, Nigeria: Northern Nigerian Publishing Company.

Marshall-Fratani, R. (1998) 'Mediating the global and local in Nigerian pentecostalism', *Journal of Religion in Africa*, **XXVIII**, 3: 278–315.

Mortimore, M. (1989) *Adapting to Drought: Farmers, Famines and Desertification in West Africa*, Cambridge: Cambridge University Press.

Mortimore, M. (1994) 'Population growth and the future dryland farming systems', West Africa Seminar, 18th March 1994, Dept. Anthropology, University College London.

Mortimore, M. J. (1968) 'Population distribution, settlement and soils in Kano Province, Northern Nigeria 1931–62' in *The Population of Tropical Africa* (eds, Caldwell, J. C. and Okonjo, C.) London: Longmans, pp. 298–306.

Mortimore, M. J. (ed.) (1970) *Zaria and its Region*, Occasional Paper no. 4: Department of Geography, Ahmadu Bello University, Zaria, Nigeria.

Munro, D. A. (1995) 'Sustainability: Rhetoric or reality?', in *A Sustainable World: Defining and Measuring Sustainable Development* (ed., Tzyna, T. C.) London: Earthscan, pp. 27–59.

Mustapha, R. A. (1991) 'Structural Adjustment and Multiple Modes of Social Livelihood in Nigeria', Discussion Paper 26, UNRISD, Geneva.

ND-HERO (1999) *News and Background on Ogoni, Shell and Nigeria* (accessed 03/02/99) www site: http://oneworld.org.delta/9906_IJAW.html.

Nigeria, Federal Office of Statistics (1995) *A Statistical Profile of Nigerian Women*, Lagos, Nigeria: Federal Office of Statistics.

Olukoshi, A. O. (ed.) (1993) *The Politics of Structural Adjustment in Nigeria*, London and Ibadan: James Currey and Heinemann.

Shenton, R. W. (1986) *The Development of Capitalism in Northern Nigeria*, London: James Currey.

Smith, J., Barau, A. D., Goldman, A. and Mareck, J. H. (1994) 'The role of technology in agricultural intensification: the evolution of maize production in the northern guinea savannah of Nigeria', *Economic Development and Cultural Change*, **42**, 537–554.

Smith, M. G. (1951) *Social and Economic Change Among Selected Native Communities in Northern Nigeria*, Ph.D thesis, University of London.

Smith, M. G. (1954 (reprinted 1981)) 'Introduction', in *Baba of Karo: A Woman* of the Muslim Hausa (ed., Smith, M. F.) London: Yale University Press, pp. 11–34.

Smith, M. G. (1955) 'The Economy of Hausa Communities of Zaria', Colonial Research Studies 16, HMSO, for the Colonial Office, London.

Smith, M. G. (1960) *Government in Zazzau*, London: Oxford University Press.

Smith, M. G. (1962) 'Exchange and marketing among the Hausa', in *Markets in Africa* (eds, Bohannon, G. and Dalton, P.) Evanston: Northwestern University Press, pp. 299–334.

Soyinka, W. (1996) *The Open Sore of a Continent*, Oxford: Oxford University Press.

Stock, R. (1985) 'The rise and fall of universal primary education in peripheral Northern Nigeria', *TESG*, **76**, 4: 274–287.

Theroux, P. (1994) 'The lepers of Moyo', *Granta*, **48**, 129–191.

Verma, O. P. and Singha, P. (1982) 'Fertility patterns of Muslim Hausa women in Northern Nigeria', *Nigerian Journal of Economic and Social Sciences*, **24**, 2: 185–198.

Watts, M. J. (1983b) *Silent Violence: Food, Famine and Peasantry in Northern Nigeria*, Berkeley: University of California Press.

World Bank (1999) *World Development Report*, Oxford: Oxford University Press.

9 Sustainability

Life chances and education in Southern Africa

Nicola Ansell

Introduction

Access to secondary education is rapidly expanding in rural Southern Africa, hence this chapter sets out to explore the role it plays in the sustainable development of rural communities. This is not a simple matter of assessing whether secondary schools teach skills which can be put into effect in raising the agricultural productivity of the land, while conserving its utility for future generations. Although such a role is doubtless relevant, and will be discussed in the course of the chapter, there are two roles played by the school which are of arguably greater significance.

First, in a region where rural communities have long served as labour reserves, the rural cannot be seen in isolation. Secondary schools play a variety of roles impinging on urban–rural interdependence, with significant implications for sustainable development. Second, as the introduction to this volume made clear, sustainability must be understood in relation to lifestyle preferences, which are themselves highly fluid. Secondary education plays a critical role in shaping lifestyle aspirations and expectations, providing young people with skills, knowledge and values relating to ways of life which differ from those of previous generations.

A further consideration of relevance to any discussion of sustainable development is the role of gender. The impacts of Southern African education are highly gendered, as are rural lifestyles and the system of rural–urban interaction. It is necessary to consider the roles commonly associated with women in relation to sustainable rural livelihoods: roles which are too frequently neglected. Of particular relevance are the performance of reproductive tasks such as collecting fuel and water; community management roles, again in relation to water and forestry resources; and the impacts of secondary education on these.

Research context

The chapter focuses on two rural communities, and the secondary schools which serve them. Ha Rateme is located in the mountains of Lesotho, four hours by bus from the capital, Maseru. Mzunga is in a Communal Area of Manicaland, a similar distance from Harare, capital of Zimbabwe. Both are villages of 40–50 households, engaging in subsistence farming, but also reliant to a large extent on the incomes of male labour migrants. Each community sends a number of children to the local secondary school. From Ha Rateme, students walk for 1½ hours to Mahloko High School, a boarding school of about 240 students. From Mzunga, students attend Ruchera Secondary School, a district council day school of 500 students.

Data was collected over two 4–5 month periods in 1996 and 1997; part of a wider programme of research. A variety of qualitative methods were used: participant observation in the two schools, interviews with community members and teachers, and focus groups of students.

Expansion of secondary education

Secondary school provision has expanded considerably throughout much of sub-Saharan Africa in the past three decades, such that by 1990 Gross Enrolment Ratios were approximately 32 per cent for boys and 15 per cent for girls (UNESCO 1990, cited in Graham-Brown 1991: 47). In Lesotho, currently about 60 per cent of girls and 39 per cent of boys begin secondary school,[1] while in Zimbabwe the figures are 64 per cent of girls and 72 per cent of boys.[2] Although the numbers in rural areas are lower, and many drop out before completing the secondary course, enthusiasm and demand for education in rural communities is substantial. Most young people of secondary school age in both Mzunga and Ha Rateme attend school. Households invest a very substantial part of their incomes in educating their young people. At MHS, fees amount to about M500 a year (US$100), with books costing a further M200 a year. RSS charges about Z$500 a year (about US$50) per student.[3] Given the time students spend in secondary school (as much as 8,000 hours over a 5-year period), secondary education must be expected to have a considerable impact on young people and their communities.

The Southern African labour reserve economy

In Southern Africa the concept of rural sustainability is problematic. The rural is difficult to define, and few rural communities can be considered in isolation from other areas to which their economies are closely tied. The colonial economy in many parts of Southern Africa relied upon rural communities as sources of labour for mines, commercial farms and urban-based production. The rural community served the capitalist economy in two ways: as the site of social reproduction of labour, and the subsidy to labour costs provided by those engaged in subsistence production, feeding the families of labour migrants (Simon 1989). In return, the earnings of male migrants from their labour in mines and industry allowed rural people to access non-rural resources, thereby extending the rural community's ecological footprint. As Figure 9.1 illustrates, the rural labour reserve and areas of commercial production (whether urban or rural) were connected through flows of labour and resources in an interdependent complex.

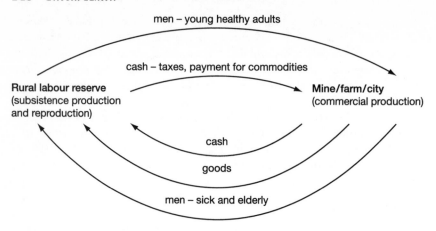

Figure 9.1 Economic integration of the labour reserve

In order to maintain this system, and provide subsidised labour for the settler economy, the independent sustainability of the rural economy was deliberately undermined, both through taxation of rural households, and by controlling land availability by removing land from peasants (successive Land Acts served this function in colonial Zimbabwe, and in Lesotho's case the annexation of the 'Conquered Territories' in 1869 removed from the Basotho the majority of their arable land). Land availability was kept at an optimum level, whereby it could be cultivated using only female labour, but was insufficient to support the rural population without additional resources, provided in exchange for male migrant labour. Commercial production by rural Africans was not viable, and, in Zimbabwe, was legally suppressed.

While this economic system may be sustainable as a whole, rural dependence and unsustainability was written into its design. Despite political independence and the demise of apartheid in South Africa, and associated ending of controls on movement, the labour migration complex continues. In 1993 40 per cent of Lesotho's GNP derived from South African mine remittances (Gay *et al.* 1995), and of men aged 20–39, 93 per cent were absent from Mzunga, and 50 per cent from Ha Rateme.

Labour migration involves not only workers in productive industries such as agriculture and mining, but also, and increasingly, service employment. Little formal employment is available in rural communities, but such work as there is, including jobs in rural clinics, schools, agriculture departments and supermarkets, in general is recruited centrally. Such employees, therefore, do not work in their 'own' communities, but participate in labour migration.

While in colonial times, however, land in rural areas was generally adequate to supplement migrant incomes, and subsidised labour in town, population pressure has since resulted in environmental degradation of what was often poor-quality land to begin with, further reducing the ability of the rural population to support its own subsistence from within the rural community. Urban areas are no longer so dependent on the rural for either reproduction of labour (which can now take place in towns, as controls over the movement of families have been lifted) or for subsistence agriculture as a subsidy to labour costs, except at times of economic crisis and retrenchment of migrant labour.

It is thus necessary to consider the role of secondary education in relation to this system: in relation to flows of labour and commodities; in relation to activities of production (both commercial and subsistence, local and remote) and reproduction (often neglected in considerations of sustainability) and their respective geographies; and in relation to the gendered organisation of the system. However, it should also be recognised that the rural community interacts with the non-local, not only in terms of the physical movement of money and people, but also in less material ways, in which education again plays a part.

Education and rural life chances

Rural livelihoods in Southern Africa are dependent upon the operation of the labour migration complex. Life chances of individuals and their households depend upon their position in this system. Direct participation in labour migration offers access to cash and goods which would not otherwise be available. In Mzunga, for instance, families in which members were in formal employment had been able to purchase televisions and install solar panels to power these, while in Ha Rateme access to migrant incomes enabled the construction of new houses and latrines.

Increasingly the employment opportunities available, even to labour migrants, are diminishing. In the past, almost any healthy Basotho man could obtain employment in the gold and diamond mines of South Africa, but entrenchment due to falling commodity prices and a post-apartheid preference for South African labour are diminishing opportunities for Basotho workers. In Zimbabwe, the economy has been in decline for some years, and unemployment is high. As employment becomes scarce, the need for formal qualifications as a pre-condition increases. Young people now have fewer employment opportunities than their parents' generation. Hence the flow of labour from rural communities is increasingly restricted to those with secondary education. Secondary education, then, is becoming an essential part of the rural reproduction of labour. However, unlike the more traditional forms of reproduction, secondary education is funded largely from outside the community, and the personnel involved in the process also originate elsewhere. It is thus an activity which cannot be sustained from within the rural community.

Rural life chances are thus closely related to education. Secondary education is perceived in many rural communities in Southern Africa as the most effective means to access the resources needed to secure household livelihoods and improve the life chances of young people. When questioned as to the reason for their investment in the secondary education of their children, parents almost invariably explain that this will enable their children to find employment (Box 9.1): that 'life, these days, depends on education'. Education is seen by parents within the context of the migrant labour complex. It is a means of participating in the flow of labour out of the rural community: a flow reciprocated by an in-flow of cash and commodities. Many parents comment that they expect their children, having obtained employment, to support the family, including by paying younger siblings' school fees.[4]

Box 9.1: interview

– Is it important to be educated?
– Yes.
– Why is it important?
– Educated people have jobs.

(Woman, aged 35, Ha Rateme, 3 May 1996.)

Given the dependence of life chances upon education, it is pertinent to explore the means by which secondary education is accessed by individuals. Given the level of fees, many poorer members of rural communities are unable to afford to send their offspring to secondary school. Such children will most likely remain in rural communities, reliant on subsistence agriculture and the wages of others.

As part of a strategy for securing rural livelihoods, secondary education is regarded by many parents as an investment in the future of both their child, and their own household. The choice to educate a particular child is often related to the likely return on the investment. Here, gender is significant. Historically most labour migrants were men. As education was perceived to be geared to participation in this system, it is unsurprising that in many Southern African communities boys were educated in preference to girls. Furthermore, girls were expected to marry and become part of another household. Other than elevating the bride-price payable for an employable daughter, sending a girl to secondary school was unlikely to help to sustain the household.

Although, in Zimbabwe, the pattern has historically conformed to expectation,[5] in Lesotho girls have long received more education than boys. This is not unrelated to the labour migration system: in the past, men could obtain mine work with little or no education. For women to obtain employment of any form, education was needed. Girls' life chances depended more on education than did those of boys. A preference for educating girls was compounded by the involvement of young boys in the rural subsistence economy: in the absence of adult men, boys were needed for herding livestock, a full-time occupation, and were hence unable to attend school. Sustenance of rural livelihoods often precluded the education of boys, even if, in the long term, individual and household life chances would have benefited.

Where parents have the luxury of choice, their decisions are frequently based on the likely returns. Increasingly, even in Zimbabwe, parents comment that girls are more likely than boys to 'remember their parents', even if married (Box 9.2), and therefore educate them accordingly.

Box 9.2: interview

– *Do you think it's more important for girls to go to school than for boys?*
– It's more important to send a girl.
– *Why's that?*
– Boys are useless. All they do is make girls pregnant and smoke dagga. A girl, when I am ill, my daughter comes with groceries.

(Woman, aged 36, Standard 6, Ha Noko, 4 July 1996.)

Often the decision to educate a particular child is made on the basis of academic ability. Students of either gender who are likely to obtain the examination certificates required by employers are worth investing in. This is due to the awareness that education in itself is insufficient to significantly enhance a child's life chances. Paper qualifications, not schooling *per se*, are arguably the most significant determinant of life chances. Once again, the factor determining life chances lies beyond the rural area.

Despite their parents' expenditure, the vast majority of rural students – over 95 per cent of those entering the first form at MHS and RSS (Table 9.1) – fail to obtain the certificates required by employers. Very few of these will obtain formal sector employment. Those students who do not obtain employment are perceived to have 'wasted' their education (Boxes 9.3 and 9.4). Parents are not unaware of the risk that investment in their children's education may not pay off. Their calculations are based on the returns which accrue to those who do succeed in obtaining employment on the basis of their education. Formal employment requiring O-level certificates yields incomes which are very much higher than those available in the informal sector.

Table 9.1 Percentage of pass rates, MHS and RSS

Year	*MHS – COSC passes*[a] *(%)*	*RSS – O level passes (%)*
1994	27.3	4
1995	10.0	5
1996	13.6	5

Source: Ministry of Education, Lesotho (1995), MHS and RSS.

Note
[a] Drop out rates at MHS are such that only 10–20% of those entering Form 1 sit O levels in Form 5.

Box 9.3: interview

I have worked hard to get money to pay for Yeukai and Norman, so if they complete their Form 4s and stay [in the village], it will be so painful, because I'll have wasted a lot of money sending them to school.

(Woman, aged 37, Mzunga, 2 October 1997.)

Box 9.4: interview

. . . Otherwise, if you don't have that secondary education, there is nothing you can do which is very productive. Otherwise, you can't be employed without O Level passes. You can't do anything. The only option is to engage in agriculture at home. That's the only option. Else you'll be a victim of circumstances, as far as the economy is concerned. That is, try to look for work, you'll be given casual labour, and then you will be given peanuts from there. Which means your life will be in ruin, or will be useless. So, secondary education is very important in the sense that, maybe, if you succeed, you will do something, and then you will be, maybe, prosperous in your life.

(Mr Nyagomo, student teacher, RSS, 23 October 1997.)

Education and rural livelihoods

Livelihood has been defined as 'adequate stocks and flows of food and cash to meet basic needs' (Elliott 1994: 64). It is appropriate to add to this the activities of social reproduction and community organisation, often largely the responsibilities of women, without which household sustenance cannot be assured. In Southern Africa members of rural communities depend for their livelihoods largely on a combination of subsistence production and migrant remittances.

A substantial part of the income of school leavers who obtain employment is fed back into rural households. Children give their parents money, which might be spent on sustaining the livelihood of the household in the form of food, or other necessities, or enabling the acquisition of material goods. Very often children support their parents, not with money, but by giving them goods (e.g. radios, televisions, clothing) which are not readily available in rural communities. Whichever form the assistance takes, it is likely to result in the importation to the rural household of goods produced outside the community. These households become increasingly dependent, not only on income derived from elsewhere, but on commodities produced elsewhere and tastes developed elsewhere.

A significant characteristic of livelihoods based on migrant remittances is their insecurity. Not only is employment frequently precarious, but migrants often find other uses for their income, so that which reaches the rural community may be sporadic. This makes many women (who are more likely to remain in the rural community[6]) very vulnerable, both to sudden loss of income, as well as to degradation of the environment which is their only other source of sustenance.

Seldom is money from employment invested in ventures which generate on-going costs, such as agriculture or other forms of rural production. Instead, consumer durables are purchased at times when income is in excess of that needed for basic survival. Hence an income generated beyond the bounds of the rural community does little to enhance rural sustainability.

The 95 per cent of school leavers without the paper qualifications to obtain employment cannot contribute significantly financially to the livelihoods of their families. They and their families are to a large extent financially dependent on the 5 per cent who are able to obtain employment. The contribution of the unemployed to sustainability must lie in the roles of reproduction, subsistence production, rural commercial production (both agricultural and non-agricultural), and community organisation. Through such activities, rural life may be supported, and rural dependency on the non-local reciprocated. However, formal secondary education does little to enable or encourage young people to engage in such activities.

Education for reproduction

Although it is not always simple to differentiate those tasks which are most appropriately defined as reproductive from those considered productive (cooking, for instance, might be deemed productive if carried out for payment, but reproductive if geared to meeting the needs of the household), much of the work carried out within the rural community is usually considered to be part of social reproduction. As such, it is most frequently performed by women. While in the colonial era, girls were educated for domestic roles in rural areas, after independence such stereotyped allocation of roles was regarded as inappropriate, and to deprive girls of the qualifications necessary for formal employment. Post-colonial education, therefore, prepares all young people, male and female for white-collar, predominantly urban employment. In neither school is any subject taught (such as home economics) which would assist students in reproductive work. Nor, beyond references to nutrition and occasionally health and sanitation, does the curriculum of either school offer any preparation for reproductive roles.

Education for agricultural production

Although agriculture is taught in both schools, it does not have a high status, and is available only to junior forms at MHS and is dropped by many at RSS. Furthermore, it is treated in textbooks[7] and lessons as a subject suitable to men, thereby alienating many girls from an activity by which many rural women are able to supplement their household incomes. Although both schools have fields which the students cultivate, as with other curricular subjects, agriculture is assessed largely by means of written examinations and the focus is more on memorisation of fact than practical application.

Although agriculture lessons (and also geography) instil in students an appreciation of the importance of the environment, and achieving sustainability (especially with regard to the need to avoid over-grazing of pasture), this is of little value unless the students are sufficiently interested, and see in agriculture a means of supporting themselves.

Education for non-agricultural production

Given the lack of formal sector employers in rural communities, opportunities for productive work outside agriculture are largely confined to self-employment and the informal sector. Secondary schools in Lesotho and Zimbabwe provide few skills for self-employment. Vocational education, other than agriculture, being more expensive than academic education, is seldom provided in rural schools. In Lesotho, the First Five Year Development Plan declared that each student would 'have sufficient practical subjects to prepare him [sic] for a less academic future' (KOL 1970: 168). Despite these sentiments, in the past 30 years, vocational education in Lesotho has seen little development.

Similarly, although in 1990 the Minister of Education in Zimbabwe announced that a quarter of school time should be devoted to specific skills training, she admitted only about 800 schools had sufficient hand tools, and only 30 secondary schools could provide higher levels of skills training (Sachikonye 1990). Syllabuses used for Commerce in Zimbabwe and Principles of Accounts in Lesotho relate little to the small businesses rural students might realistically establish.

Education for community organisation

Community organisation is vital if the environment is to be managed in such a way as to maintain its sustainability. This is frequently a task for which women are largely responsible. However, preparation for community organisation does not figure highly in schools' priorities. Students leave school with limited knowledge of community structures (particularly those of a democratic nature). Furthermore, the relationships between schools and their local communities are commonly tense, with communities perceiving schools as alien intrusions, controlled from elsewhere (although at times members of school committees endeavour to exercise power over teachers). Teachers at Mahloko High School, for instance, complained the local people considered them 'ex-pats'. At neither school did teachers consider themselves part of the local community, or feel they had any obligation towards or any involvement with the community outside their immediate employment. Students too, especially at MHS, learn to see themselves as outsiders to the community.

Furthermore, students are poorly equipped by schooling to make decisions, or solve problems. In secondary school, knowledge is presented as derived from elsewhere. Creativity is seldom encouraged. Hence it is not surprising women seek answers to their problems from elsewhere, rather than finding creative solutions from within the community.

In general, then, schools fail to provide students with skills or knowledge relevant to productive life in a rural community, thereby failing to enhance the sustainability of rural livelihoods.

Education and lifestyle aspirations

Not only does education impinge on students' abilities in the productive and reproductive spheres (their capacity to sustain a livelihood); it also affects aspirations concerning lifestyle (consumption preferences), and in this way, too, is of significance to discussions of sustainability.

Clearly the distinction between livelihood and lifestyle is not this clear-cut. All rural people aspire to lifestyles in which their livelihoods are secure (this is generally the first priority of those earning an income); and what constitutes a livelihood is, to some extent, a question of expectation. However, it is not unreasonable to draw a distinction between that which is essential to sustain life, and consumption patterns relating more to style and fashion.

Education is seen by those involved in its planning, delivery and consumption, to be geared primarily to suiting students to employment. Given that employment is not generally attainable in the local community, this implies that education is geared to the labour migration complex. This is particularly strongly reinforced through the focus of secondary education on preparation for examinations leading to paper qualifications, which are of little value other than in the formal employment market. Although most students fail to obtain these, this merely persuades them that their education has been wasted: not that it could be seen to serve an alternative purpose. Hence the entire education system appears to be gained to obtaining employment which lies beyond the immediate rural community. Any income earning potential acquired in rural schools does not equip students to engage in productive work within their own communities. Rather, 'successful' secondary school graduates are those who find employment elsewhere, often in urban areas.

The focus on examinations contributes to and compounds the effects of the lack of attention paid to informal sector/self-employment. Most young people in Lesotho do not consider employment in the informal sector to be employment (Phororo *et al.* 1993), and manual labour in Zimbabwe is held in low esteem (a result in part of the academic emphasis of education (Chigwedere n.d.)). Only one boy and no girls at Mahloko High School, for instance, wanted to be self-employed.[8] Similarly, the absence of reproductive work from the curriculum serves to diminish the status attached to such work in the eyes of students. Students desire (and in many cases expect) formal sector work and the income it will bring. They learn to think in terms of a future lifestyle in line with those available to people employed in white-collar jobs: a large house, television, possibly a motor vehicle, with no reason to labour on the fields.

Although they desire formal sector employment, students do not always wish to live in an urban environment. Indeed many list the advantages to rural dwelling, and would prefer to remain in a rural environment. They are aware, however, that these wishes are incompatible. Even if they obtain rural employment, for example as a nurse or teacher, it is highly unlikely to be in their 'home' community. As an employee they would not 'belong' to the community in which they worked, and would not expect to contribute to that community, other than in the context of their employment. Schools do not encourage students to consider active involvement in the community in which they work. This may be partly attributable to lack of role models, teachers having little involvement in the communities local to the schools. Hence those who obtain formal employment are unlikely to work in their communities for sustainable development.

Although students may express a desire to help their 'home' communities, unless they live in the village, they are unlikely to participate in rural community life. Since students do not aspire to remain living in their home community (in which they would be unemployed), they seldom think in terms of how they might participate in community life in this, their most likely role.

Through their focus on employment, and the examinations which need to be passed to achieve this, schools encourage in students aspirations which can only be met outside the immediate rural context. Both the employment itself, and the income needed to support the desired lifestyle can only be sourced elsewhere. Hence secondary school leavers' desired lifestyles cannot be considered sustainable within a rural environment.

Conclusions

This chapter has addressed the impact of secondary schools on rural life chances: in particular the question of whether secondary education enhances the abilities of people to sustain a rural livelihood, or the lifestyle to which they aspire. Rural life chances depend to a large extent on being able to obtain formal employment outside the immediate rural area: upon participating in a labour migration complex as a labour migrant. The opportunity to gain such employment in turn depends upon obtaining secondary education, but it is insufficient simply to attend secondary school: it is necessary to gain appropriate formal qualifications. Clearly, secondary education gives those individuals who obtain such qualifications the ability to sustain their desired lifestyle, and, in a few cases, to sustain the livelihoods of extended rural households. However, such individuals and households are almost always dependent on income sourced outside the immediate rural area – even, in the case of many Basotho, from beyond the national borders, thus extending the ecological footprint. Furthermore, they draw upon resources which may themselves be unsustainable, particularly given the current economic climate in Southern Africa.

Many, however, do not acquire such an ability to source their chosen lifestyle from beyond the rural community, but are compelled to support themselves within the local community. The rural secondary school does not prepare students for such a lifestyle, and, beyond aspiring to smaller families than their parents' generation, they are little prepared, either in terms of farming skills, or community management skills, to pursue rural development in a sustainable manner.

The majority of school leavers remain living in rural areas, unable to access the resources which would provide them with the lifestyles they desire. They also lack the capacity to secure the livelihoods of their families from within the rural environment. Rather, they are increasingly dependent on that minority of the community who are able to participate in the 'modern' sector of the economy and contribute to meeting the lifestyle and livelihood demands of the rural population. Hence, secondary education contributes to the entrenchment of the (gendered) labour migration complex, which undermines rural sustainability, enforcing dependence on the non-local.

This raises the question, as yet unconsidered, of whether such interdependence is necessarily problematic. Rural communities cannot be expected to be entirely self-sufficient, especially given their history. If the system is itself sustainable, it may be the best means of securing rural livelihoods in the long term. Nor do aspirations to formal employment and urban lifestyles necessarily mean urbanisation and unsustainability. It is argued that peri-urban areas of Lesotho make the most productive use of land: whereas 'rural' land is farmed extensively for maize and over-grazed by livestock, peri-urban land is farmed intensively as vegetable gardens, with careful cultivation, and regular application of water and manure (Gay *et al.* 1995), while benefiting from access to employment sources.

The labour migration complex is not static, but constantly evolving. The question remains as to whether it is sustainable. Simon (1989: 162) comments: 'At present, survival for most worker-peasants in southern Africa depends on maintaining footholds, however precarious, in *both* rural and urban areas.' However, it is unlikely that those with urban employment any longer *need* their rural base. No longer is the non-rural economy so dependent on the rural. Subsistence agricultural production has declined, particularly in Lesotho, and the non-rural economy no longer requires so much labour. Increasingly, jobs in urban areas can be filled with labour which is supported and reproduced within the urban environment. However, the rural is increasingly dependent on the urban. Secondary education plays a key role in sustaining this dependency and hence undermining the sustainability of rural life.

For rural livelihoods to be sustainable, rural areas need to provide opportunities for people to spend their time in activities which will either support rural livelihoods directly, or will increase the reliability of resource transfers from outside the community. These must be both reliable and equitable. Currently a tiny percentage stand to gain highly, but with considerable uncertainty, whilst the vast majority of rural people remain dependent upon them.

Sustainable rural livelihoods thus demand a different form of education: one in which obtaining employment beyond the community was not the sole measure of success; in which skills of use in rural communities, including both productive, reproductive and community organisational skills were valued; in which students could expect to live in a rural community in which they would have a personal interest and role to play; in which the only knowledge valued did not derive from, and get measured, outside the immediate environment.

However, schools are not simply manipulable tools which can be used to provide a technical 'fix', but are an integral part of the way in which society is organised and power exercised. Education does not exist in a vacuum, but is the result of policies and practices, past and present, national and international. Education systems in Southern Africa are generally highly centralised, and schools are designed to serve particular interests: interests which are often urban-based. Education is seen, even more than in colonial times, as a means of producing a workforce.

Ultimately the current form of rural secondary education is probably not sustainable insofar as large amounts of both rural and government money are invested in schools which produce only a very small number of productive workers, and a much greater number of dependents. Education provides only limited abilities to sustain rural livelihoods, but aspirations to different lifestyles which cannot be sustained from within a rural environment.

Notes

1   Calculated from statistics on school enrolment (1993) (Ministry of Education 1993) and population projections based on census data (BOS 1997).
2   Calculated from statistics on school enrolment (1995) (CSO 1997) and population projections based on census data (CSO 1994).
3   These figures compare with 1997 per capita GNP in Lesotho of US$670 and in Zimbabwe of $750 (World Bank 1999).
4   Not all parents expect their children to contribute to the household. Some (particularly those who are more financially secure) are primarily concerned with their children's own life chances, remarking that they want them to 'have a better future' or 'not to have to struggle like I have'.
5   This pattern persists: while 63 per cent of Zimbabwean boys complete their secondary education, only 48 per cent of girls achieve this: girls constitute only 43 per cent of students at Ruchera Secondary School.
6   Men often migrate in search of work, even if they lack qualifications.
7   For instance, one of the agriculture textbooks used at Ruchera Secondary School (Owen 1984) contains 33 illustrations portraying people: of these, despite the fact that the majority of farmers in Zimbabwe are women, only one picture clearly portrays women or girls, while two others picture people of indeterminate gender. In general, textbooks only show women engaged in manual and stereotypically female tasks. Technology and livestock are monopolised by men.
8   In the event the boy passed COSC and decided to study further.

References

BOS (1997) 'Statistics on gender in Lesotho.' (Bureau of Statistics, Government of Lesotho, Maseru) mimeo June 1993.

Chigwedere, A. (n.d.) *The abandoned adolescents.* (Mutapa, Marondera).

CSO (1994) *Census 1992: Zimbabwe National Report.* (Central Statistical Office, Harare).

CSO (1997) 'Education statistics.' (Central Statistical Office, Harare) mimeo.

Elliot, J.A. (1994) *An introduction to sustainable development.* (Routledge, London).

Gay, J., Gill, D. and Hall, D. (1995) *Lesotho's long journey: hard choices at the cross-roads: a comprehensive overview of Lesotho's historical, social, economic and political development with a view to the future.* (Sechaba Consultants, Maseru, Lesotho).

Graham-Brown, S. (1991) *Education in the developing world: conflict and crisis.* (Longman, London and New York).

KOL (1970) *Lesotho First Five Year Plan 1970/71 – 1974/75.* (Central Planning and Development Office, Maseru).

Ministry of Education (1993) 'Education Statistics.' (Planning Unit, Maseru).

Ministry of Education (1995) 'An analysis of students' performance in COSC since 1990.' (GOL, Maseru).

Owen, G. (1984) *O-Level agriculture.* (Longman, Harlow).

Phororo, H., Mapetla, M. and Prasad, G. (1993) 'Youth aspirations and needs in Lesotho.' (ISAS, Rome).

Sachikonye, L. (1990) 'Strains and stresses in the Zimbabwean education system: an interview with Fay Chung, Minister of Education and Culture.' *ROAPE* 48: 76–81.

Simon, D. (1989) 'Rural–urban interaction and development in Southern Africa: the implications of reduced labour migration.' in Potter, R.B. and Unwin, T. (eds) *The geography of urban–rural interaction in developing countries.* (Routledge, London).

World Bank (1999) *World development report: knowledge for development.* (Oxford University Press, New York).

10 Linking the past with the future

Maintaining livelihood strategies for indigenous forest dwellers in Guyana

Caroline Sullivan

Introduction

Amerindian livelihoods demonstrate a depth of ecological knowledge rarely found in other societies. Thousands of years of survival strategies have provided the social capital needed to generate the multi-sectoral income flows found today in Amerindian communities. Renewable, organic materials from the forest eco-system provided the base for almost every facet of life, while traditional management practices and the low density of human populations ensured that the carrying capacity of the system was never reached.

Today these lifestyles still provide the basis for everyday living, but economic and social pressures are inevitably bringing changes to the traditional ways of life. In the face of mounting pressures of economic globalisation, resource extraction from Guyanese forests is increasing at an alarming rate, inevitably having serious impacts on Amerindian communities. In the spirit of Our Common Future (WCED 1987), and Agenda 21 (UNCED 1992), the challenge today is to use the whole spectrum of our knowledge to develop new paradigms of thought through which sustainability can be operationalised, and livelihoods such as these may be maintained.

People in the forests of North-West Guyana

Archeological records (Bruhns 1994) suggest that ancient, organised societies existed in the western part of Guyana as long ago as 3000 BC, and evidence suggests that they remained there in growing numbers up until 900 BC. Ceramics of this early period (known as the Alaka period) are functional and rounded, often strengthened with crushed shells, and decorated with a red wash made from local lateritic clay. Later ceramic evidence suggests that by 100 BC, during the period referred to as the Mabaruma phase, which lasted until about AD 1400, communities had become more sophisticated, and produced decorated ceramics which have clear stylistic links to the Los Barrancos and La Gruta pottery found in the Orinoco Basin.

Artifacts found at these archaeological sites suggest that these were agricultural communities which depended on Cassava (Manioc) as their staple food, and capitalised extensively on fish from the richly stocked rivers and flooded savannah. Villages evolved on raised levees, naturally formed from alluvial deposits, and covered by sand blown up from the river beds during the dry season. These islands in the semi-permanent wetlands of the savannah provided ideal places for the establishment of hunting camps, which later developed into settlements with complex political and social structures. According to early Dutch colonial administrators in the eighteenth century, these communities were described as the 'free nations', and were largely made up of people from the Carib, Arawak, Akawaio, and Warau tribes (Forte 1995).

Today, in North-West Guyana, many Amerindian settlements are still located on these raised sand levees, with their lifestyle still based almost exclusively on the exploitation of environmental goods and services from the forests. Currently, the estimated population of Amerindians living in the North-West district of Guyana is 14,075 (GOG 1993), and while many of these people are descendants of tribal groups which migrated from Venezuelan territory during the nineteenth century, people in Amerindian communities throughout the region have a relatively homogenous lifestyle and socio-economic profile, and, in monetary terms, rank amongst the poorest in the world.

In spite of the extensive changes which have taken place elsewhere, in reality, little has changed for Amerindian people over thousands of years. Households today are still dependent on the same ecological cycles of seasonal flooding and nutrient release as their ancestors had been, with domestic skills of weaving and basketry still maintained and used to the same ends. People are still living mainly as farmers, depending on cassava as their staple food, supplemented by fish from the flooded swamp, and animals and birds hunted from the forests. Again, archaeological evidence suggests that birds and monkeys have always been important to the diet of these forest people (Bruhns 1994), and in the case of forest birds, this is still observable today. This tends to suggest that the traditional way of life of the Amerindian people of this region is one which, to a large extent at least, can be said to have been economically, socially and ecologically sustainable for thousands of years, and as such, its value must be recognised.

The indigenous knowledge held by these forest dwelling people was first recognised in the West by Sir Walter Raleigh, who, in his search for the famed gold of El Dorado, was convinced that it was to be found under a lake in Guyana (Hills 1961). This resulted in numerous expeditions to the interior, and in his reports of 1595, he mentions the use of Curare, a toxic plant substance prepared by Amerindians for use in hunting. It was not until the late eighteenth century that even a preliminary chemical analysis of this was made, and even today, modern chemical techniques still cannot fully determine the exact composition of the poison (DeFillips 1992). This is true of many of the medicinal plants which are in use by Amerindians in Guyana today, and it reflects the fact that knowledge of plant utilisation in Guyanese forests is something which is deeply embedded in the culture of Amerindian people.

Another example of this is the collection of non-timber forest products for food and drink, which is very widespread in these communities. According to data collected from Amerindian households in 1996, 39 per cent of women regularly collect food from the forest, while 38 per cent collect medicinal plants, and 21 per cent collect handicraft materials (Sullivan 1997). In addition to the items collected for the household by women, almost everyone in these Amerindian communities collects fruits and nuts opportunistically whilst in the forest. This collection of forest foods provides an important degree of food security for Amerindians, since this type of wild food is especially important in times of crop failure, or other hardship, and in addition, the nutritional value of such foods, although not well known, has been shown to be high in vitamins and minerals essential to good health (Melnyck 1995).

Traditional Amerindian farming techniques, involving inter-cropping with as many as twenty-five different plants species, also illustrates the type of complex ecological knowledge held by Amerindian communities (Forte 1996a). While the traditional process of slash-and-burn farming has been criticised in recent years, there is no doubt that the technique is both effective and sustainable for small stable populations in forest areas, and it is really rapid population growth which has given rise to widespread deforestation, rather than the technique itself.

One of the most striking examples of traditional Amerindian knowledge is that relating to the processing of Bitter Cassava, the basic staple food. In order to extract lethal poison contained in this drought-resistant root-crop, it needs to be processed by means of closely woven squeezers made from Mucru plants collected from the forest. This is reflected in the fact that when a sample of Amerindian villagers were questioned in 1996, 100 per cent of both men and women considered plants from the forest to be essential to life (Sullivan 1997). From this, it can be inferred that if the Amerindian lifestyle is to be secured, the concept of sustainable development must include views of Amerindians themselves, about what this actually means for them. Since to date Amerindians have had little economic or political influence in the wider community, their lifestyle is presently under threat, mainly from economic encroachment on the forest ecosystem on which they depend.

Encroachment on the economic frontiers of Amerindian lifestyles

Encroachment on Amerindian lifestyles has taken many forms over the years. Historical post-Colombian geographical encroachment, with its consequent encroachment on human health, devastated Amerindian populations, while intellectual encroachment through the spread of religion weakened traditional belief systems. Since colonial times, political encroachment through the introduction of the mission system limited land tenure, and today, economic encroachment threatens lifestyles both through depletion of the ecosystem by commercial timber extraction, and through the legislative supremacy of the laws of Guyana relating to mineral rights.

The dilemma faced by the government of any country rich in forest resources is well illustrated by the case of one species, Greenheart (*Chlorocardium rodiei*). Of the many thousands of different living species found within the Guyana forests, this is considered the most valuable of trees extracted, due to its great strength, and chemical constituents which make it resistant to marine water corrosion. These qualities make it a very important target for logging interests, but in addition to its use as an important hardwood, it is known (Fanshawe 1948) that it contains two alkaloids, Birbirine and Sipirine, which have potential for use as a treatment for malaria and other illnesses. This means that the option value which it holds may be much more important to humans than its current use value as timber, and excessive harvesting should thus be avoided as being economically unsustainable. In addition, such harvesting may be ecologically unsustainable, due to the longevity of the species lifetime, which means that it may take 400 years to reach a size consistent with extraction (Pons 1996).

The use of barks, leaves, extracts, etc. for medicine also applies to many other hardwoods in the Guyanese forests, such as Mora (*Mora excelsa*), Ubudi (*Anacardium giganteum*), and Crabwood (*Carapa guianensis*), and so the option values of these may also be very high. A total of over 127 plants from these forests were listed by Fanshawe (1948) as having medicinal properties, and many more have come to light since then, as a result of ethnographic studies conducted with a variety of tribal groups in the Guyanese interior. Given the potential revenues which could be generated from such medicinal use of plants (Mendelsohn and Balik 1995), it is important for policy-makers to consider this when allocating forest areas as logging concessions.

As originally indicated by the reports of Sir Walter Raleigh, in addition to logging activities in the forest, mining operations have always been important to Guyana, with gold and diamonds being found in large quantities in the mountainous areas of the interior, and in the catchment areas of the Mazaruni and Potaro rivers. Since 1884, gold has been extracted on a commercial basis from many inland parts of the country, and this has had a significant effect on Amerindians in a number of ways, not least being the disruption to traditional life which such employment often brings (Forte 1996). The official figures for gold production for many years remained fairly constant, at less than 1000 kg per year until 1990, when production started to rise and reached 11,678 kg by 1994 (GOG 1995). This reflected the increasing scale of mining operations, and the fact that in the past, large quantities of both gold and diamonds were smuggled out of the country to avoid tax payments. With a land border of 2,462 km, and a coast of 459 km, this problem was difficult to control, and for the Government, one of the advantages of awarding mining rights to larger operators is that it may be easier to monitor production. Since the 1970s, this loss of national earnings through informal economic activity is another contributory factor to the low levels of national economic growth in Guyana, which in itself has contributed to the economic threat to the Amerindian way of life.

Gold extraction activities include land mining and river dredging, and in many areas these have led to severe environmental damage. A recent example of such damage is provided by the Omai cyanide spill (WRM 1997), which occurred as a result of a breach in a tailings dam at a large-scale gold mining operation, in August 1995. Following this breach, 3.2 billion litres of cyanide and heavy metal liquid wastes escaped, killing all aquatic life in the Omai river, and polluting hundreds of miles of the Essequibo river, one of the major drainage systems of the Guiana shield. This has been described as one of the worst environmental disasters in South America, and in addition to the very serious ecological consequences, many local indigenous groups were affected by this spill, and their way of life severely threatened. The degree of damage sustained has been highlighted by a law suit which was filed in the Quebec Superior Court in March 1997, against the parent company, Cambior. In addition, at the International Peoples Tribunal, in New York in June 1997, individual petitions were made by representatives of the Amerindian settlements along the river, but to date, little has been, or is being, done to remedy the situation, or pay damages to the people (Gaia Forest Conservation Archive 1998).

In addition to the existing large number of small and medium scale mining operations which are ongoing in Guyana, a number of major multi-national mining companies have been granted concessions to exploit mineral wealth in the interior. Companies such as Golden Star Resources, and Zamuteba of Brazil (both of which have dubious environmental records) have recently been granted large concessions (WRM 1997), often overlapping timber sales agreements already awarded. In the case of Zamuteba, a total area of 1.7 million ha has been allocated for geological and geophysical surveys for all minerals, and in many places these concessions include areas designated as Amerindian Reservations. According to the laws of Guyana (GOG 1989), all sub-soil materials[1] are the property of the state (Flaming *et al.* 1995), and mining rights take precedence over all other legislation (except in the case of Timber Sales Agreements). As a result, there is no protection in law for the Amerindian communities, and in future it is possible that villages may become the focus of mining operations, with all the social and ecological disruption which this may bring.

In the case of Golden Star Resources and the Barama Timber Company, discussions have already been started to develop an integrated strategy of mineral and timber exploitation, which will be mutually beneficial to the two companies. Although this may make good economic and business sense, it is unlikely to be beneficial to the environment, the local population, or the nation as a whole.[2] Another example of a large mining operator which has signed an agreement on mineral exploration in Guyana in 1996, is Broken Hill Proprietary (BHP) of Australia, infamous for its ill-treatment of indigenous people in Australia and New Guinea (Colchester 1993). In the case of this company, the total size of the concession is close to 5 per cent of the country, and it seems that at least some degree of conflict of interest between the company and the local populations is likely to occur.

A case study of Amerindian lifestyles in North-West Guyana

Valuations of resources in tropical forest ecosystems often fail to take account of the full spectrum of forest products and services, since many of these have traditionally been ignored as being insignificant or non-marketed. Methodologies currently used for environmental valuation are often inappropriate in developing countries, and subject to serious degrees of error which inevitably have important policy implications (Markandya and Perrings 1993). In particular, resource auditing procedures, which may be useful in other circumstances, fail in this case because of such problems as the lack of clearly defined markets, uncertainty regarding both current and future demand and supply of forest products, and the lack of detailed information about how these resources are used (Daly 1989).

As a means of overcoming some of these difficulties, this case study attempts to examine the use-value for non-timber forest resources on the basis of how they are currently used by indigenous forest dwellers. This has been done by collecting a wide range of both qualitative and quantitative data in three Amerindian villages in Guyana, and on the basis of this data, an income accounting framework which models these village economies has been developed. Through this framework, the actual use-values of these non-timber products has been derived in monetary terms.

Funded by the Tropenbos Foundation of the Netherlands, the fieldwork for this study was conducted in three Amerindian villages in Guyana. Data was collected in a participatory manner with the help of seven Amerindian field assistants. Some broad characteristics of the study sites are shown in Table 10.1, with their geographical location shown in Figure 10.1. A number of forest-based activities contribute to the economic profile of these Amerindian households, and in addition to farming, these include the collection of fuelwood, food and drink, roofing and handicraft materials, along with wildlife and fish. Detailed data was collected from a total of 143 households, and a statistical 'snap-shot' of the economy of the villages was developed.[3]

In addition to the analysis of the monetary values of forest use, it has also been possible to examine indigenous social values of forests, and from this it can be seen that in addition to the significant role played by the forest in the economy of Amerindian villages, its importance goes well beyond the monetary sphere. It is inevitable that if such values are not included in the decision-making process, policy failures will result, and traditional ways of life will be lost.

In these villages, all farms use a system of inter-cropping, in which a large number of different crops are grown in each field. This is the typical pattern found in forest communities such as these (Forte 1995), and Tables 10.2 and 10.3 show crop distribution in the study villages.

It is interesting to note the significant differences which exist between the villages in terms of production of specific crops. These differences are the result of variation in household tastes (as exemplified by the large amount of sweet cassava being grown in Assakata), local needs (as with cotton in Karaburi where

Table 10.1 Characteristics of study villages

Attribute	Assakata	Sebai	Karaburi
Population	176	201	559
Number of households	23	31	89
Average number in household	7.6 persons	6.5	6.8
Average age of household heads	45 years	42.5	44.5
Average age of senior women in households	38 years	36.6	39.4
Average number of years education for adults	4.03 years	6.1 years	6.53 years
Average size of farm	3.4 acres	3.43	5.64 acres
Location	Assakata Creek, Santa Rosa/ Moruka	Sebai Creek, Port Kaituma	Kumaka Road, Santa Rosa/ Moruka
Vegetation type	Mixed deciduous forest and swamp, rainfall average of about 120 inches per year	Mixed deciduous forest and swamp, rainfall average of about 120 inches per year	Mixed deciduous forest and swamp, rainfall average of about 120 inches per year
Means of transport and hours of travel from the nearest market	6 hours paddling by canoe	6 hours paddling by canoe	6 hours walking, or by canoe for those households near river

craftworkers use this for making hammocks), and geographical factors such as soil quality and water supply (as demonstrated by the large output of watermelon in Assakata, located in a wetland area).

Marketing conditions are another very significant influence on farm output, and this is reflected in the price differentials[4] to be found in these villages, illustrated in Table 10.4. In Assakata, the nature of production and trade is very localised, with lower prices, and where on average 87.5 per cent of output is used at home, and only 12.5 per cent is sold. This contrasts with the figures for the other villages, where different conditions exist. In Karaburi, a village developed for agricultural production, 73 per cent of farm output is used at home, with only 27 per cent being sold, reflecting the difficulty which people here face with transportation of their crops. By contrast, in Sebai, where although the nearest market is equally far away, it is accessible by river, and a strong demand exists due to the relatively large population of timber workers living in that town. As a result, in this village, only 59 per cent of farm output is used at home, while 41 per cent is sold or exchanged.

Comparisons in labour supply and sectoral output

The variation in the quantity of labour supplied by households in the villages tends to support the idea of farmers in Karaburi having the benefit of good soil,

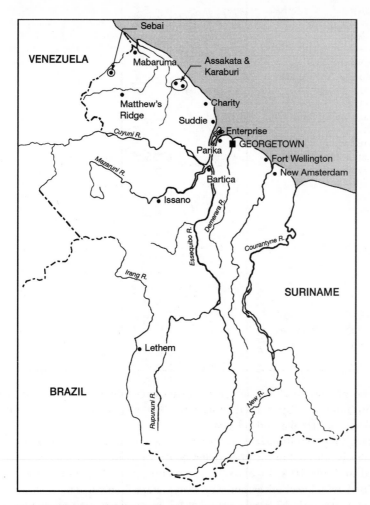

Figure 10.1 The location of the study villages in North-West Guyana

yet not being motivated to work it to its full potential. As can be seen from Table 10.5, total household hours of labour is lowest in Karaburi, and highest in Assakata, with households in Sebai falling between the two. This data suggests that households with better environmental or economic endowments such as good soil (Karaburi), or strong market (Sebai), will need to put in less labour than those where natural and economic conditions are less favourable (Assakata). This highlights the fact that within the classification of 'subsistence agriculture', there is a degree of heterogeneity amongst households. In the examples of these study villages, this is demonstrated by the fact that households in Assakata on average put in 48 per cent more labour than do those in Karaburi, and 9 per cent more than those in Sebai, while at the same time, producing less.

How the labour is used by households in the different villages is shown in Table

Table 10.2 A comparison of total crop output (in lbs) in three Amerindian villages, Guyana, 1996 (1)

Village	Bitter cassava	Sweet cassava	Yams	Eddoes	Bananas	Plantains	Sugar cane	Peppers	Pineapples	Pumpkins	Watermelons	Sweet potatoes	Beans	Mangoes	Peanuts	Other fruit
Assakata	166,127	6,250	15,624	10,068	40,743	8,884	8,518	2,061	11,776	2,096	10,168	4,814	0	0	0	554
Sebai	301,212	714	35,642	18,482	151,229	119,642	27,333	620	10,363	1,210	1,367	17,880	590	0	1,280	500
Karaburi	737,428	2,286	40,652	21,403	136,217	99,989	9,563	4,498	5,440	3,079	62	156	6,518	8,783	49,900	2,876

Note
Zeros indicate no or negligible output.

Table 10.3 A comparison of total crop output (in lbs) in three Amerindian villages, Guyana, 1996 (2)

Village	Medicinal plants	Culinary herbs	Greens	Tania	Coconuts	Irish potatoes	Coffee	Other citrus	Cotton	Maize/corn	Avocados	Squash	Cucumbers	Tomatoes	Eggs	Pigs	Total for all crops (lbs)
Assakata	148	0	9,350	3,305	0	80	330	282	0	10,270	1,145	1,037	5,231	0	20,448	0	339,309
Sebai	260	4,310	6,779	3,994	0	604	221	0	0	18,137	100	355	5,390	643	69,408	0	788,761
Karaburi	100	182	17,839	10,135	47,932	6,163	1,072	6,947	1,516	2,004	3,275	1,368	4,751	888	116,928	40	1,349,990

Note
Zeros indicate no or negligible output.

Table 10.4 Crop prices in Assakata, Sebai and Karaburi, 1996 (G$)

Village	Bitter cassava	Sweet cassava	Yams	Eddoes	Bananas	Plantains	Sugar cane	Peppers	Pineapples	Pumpkins	Watermelons	Sweet potatoes	Beans	Mangoes	Peanuts	Other fruit
Assakata	5	15	15	15	13	17	5	103	12	13	13	15	105	0	0	60
Sebai	7	17	18	17	14	17	16	68	57	18	67	17	68	0	100	50
Karaburi	7	23	22	17	16	18	18	43	24	22	20	17	69	30	84	75

Village	Medicinal plants	Culinary herbs	Greens	Tania	Coconuts	Irish potatoes	Coffee	Other citrus	Cotton	Maize/corn	Avocados	Squash	Cucumbers	Tomatoes	Eggs	Pigs
Assakata	100	0	100	15	0	15	150	60	0	35	100	50	30	0	25	0
Sebai	100	38	24	17	0	13	150	0	0	19	30	14	19	85	25	0
Karaburi	100	88	51	17	30	22	126	50	100	48	50	14	20	35	25	1,000

Note
Prices shown are per pound, except for eggs, per item, and pigs, per animal.

Table 10.5 Efffective labour supply (in hours per year)

	Assakata	Sebai	Karaburi
Total labour supply (hrs/yr)	149,430	184,810	389,277
Average labour supply by household (hrs/yr)	6,497	5,962	4,374

10.6. By examining the average values of labour by sector,[5] it can be seen that for the villages of Assakata and Sebai, the amount of effort spent in hunting and fishing is far higher than in the village of Karaburi. This indicates that for those in Assakata, where farming is less productive, significant time is spent in other forest-based activities, such as hunting, fishing and palm-heart harvesting, these being the means by which the household income can be supplemented. Furthermore, the relatively large number of hours per household spent collecting non-timber forest products from the forest (especially in Assakata), indicates the important role played by the forest in the survival strategies of the poorest forest households.

The analysis of labour utilisation in different activities within the villages does suggest that households respond differently to the various constraints that they face. In most cases, in addition to the level of labour used, other factors such as soil quality, motivation, price etc., may have an influence on the observed variation in the levels of household output. In the case of prices, as shown in Table 10.7, significant variations in prices for fish and bushmeat do exist between the villages, and as shown in Table 10.8, this certainly influences the total value of production in those sectors.

These figures demonstrate the very important role that wildlife plays in these forest households. In addition to important food animals such as Agouti and Labba frequently caught by hunters in all villages, it also reflects the large numbers of parrots and macaws caught by trappers to meet the demand from the wildlife trade. For households in both Karaburi and Sebai, trade in animals and birds is a less important activity, as other household opportunities (such as farming or handicrafts) are available.

A closer look at how these output values are distributed within the villages is provided in Figure 10.2, and again this highlights the variations which exist between apparently homogenous households in the face of variable environmental endowments and endogenous economic constraints.

The use of forest inputs in the production of household output

In addition to the use of labour and capital to produce output, households in these villages clearly make use of a significant amount of forest inputs in the production process. This plant use provides a wide variety of foods, drinks, fibres, resins and other chemicals important to the Amerindian lifestyle. One example of how such materials are used is for medicinal purposes. Of those surveyed in 1996, 52 per cent of households were found to collect medicinal plants on a regular basis. These are used to treat ailments like malaria and snake-bite, as well as other

Table 10.6 A comparison of total and average labour input values, by sector, in three Amerindian villages in Guyana, 1996 (G$)

Village	Farming	Fuelwood collection	NTFP collection	Fishing	Handicrafts	Annual palm-heart harvesting hours	Hunting	Annual value (G$)
Total								
Assakata	3,972,837	878,445	273,185	1,777,096	1,060,290	1,041,695	1,456,560	10,460,108
Sebai	5,625,963	1,008,315	141,278	2,527,560	642,600	96,600	2,894,378	12,936,693
Karaburi	15,697,178	3,011,252	413,859	2,843,773	3,515,400	199,962	1,567,944	27,249,367
Average								
Assakata	173,732	38,193	11,878	77,265	46,100	45,291	63,329	454,787
Sebai	181,483	32,526	4,557	81,534	20,729	3,116	93,367	417,313
Karaburi	176,373	33,834	4,650	31,953	39,499	2,247	17,617	306,173

Table 10.7 Meat and fish price variation across villages, 1996

	Assakata	Sebai	Karaburi
Average price of bushmeat (G$)	62.6	103.1	100.6
Average price of fish (G$)	54	104.7	76.7

frequently occurring illnesses such as fever, colds, diarrhoea and dysentery. Although conventional treatment for these is freely available from village health workers, householders often prefer to treat themselves using medicinal plants from the forest, and on average, women in these villages knew of, and regularly used, over eight different medicinal plants.

The sale of forest products outside of the household represents one of the first links to the global economy. Taking the plants collected from the forest by these households, but not including roofing materials or professional palm-heart harvesting, 22 per cent are sold or exchanged, while 78 per cent are used at home. In addition to plant use, the consumption of both animals and fish is very important for households such as those in these villages, as it is from these that the majority of dietary protein is obtained (FAO 1989).

The widespread use of medicinal plants in these villages once again highlights the importance of non-timber products to forest-dwelling people, and the failure to take account of these values is one of the factors which has given rise to some of the market and institutional failures found in these situations. One way to overcome these is to try to incorporate some measure of the value of all forest products into national accounting structures. This would ensure that any loss of such resources is compensated for in the pricing structure of forest use, thus removing externalities, and creating a more sustainable 'pareto optimum' situation of resource use (Pearce and Turner 1990).

Wildlife use in Amerindian villages

A close examination of the hunting habits in these villages reveals that agouti, labba, deer, tapir and wild hog are all considered important, while the macaw and parrot are important for hunters and trappers involved in the wildlife trade. The monetary value of hunting and fishing catches clearly demonstrates the contribution that these activities make to household well-being. Such sales provide much needed cash income for those participating households, and this is usually used to purchase items such as kerosene, tools, candles, soap and clothing.

By looking more closely at the consumption of both meat and fish within the villages, by weight, rather than by value, it is interesting to note that taken across the total population of the three villages, the average per capita weight of bushmeat consumed is 420 pounds per year, while fish consumption on average is 128 pounds per capita annually. Details of this are shown in Tables 10.9 and 10.10, and it is clear that total output of bushmeat is highest in Sebai, and the fact that a relatively smaller proportion of this catch is retained for home use reflects the

Table 10.8 Productive sectoral output values for three Amerindian villages in Guyana, 1996 (G$)

Village	Farming	Fuelwood collection	NTFP collection	Fishing	Handicrafts	Palm-heart	Hunting	Total value of outputs (G$)
Assakata	4,893,726	1,054,134	446,948	2,822,269	552,500	1,348,970	4,687,466	15,806,013
Sebai	12,907,689	1,209,978	775,909	6,766,606	965,000	92,000	10,465,432	33,182,613
Karaburi	26,058,800	3,613,502	1,051,261	5,709,020	4,746,500	217,350	7,808,030	49,168,462

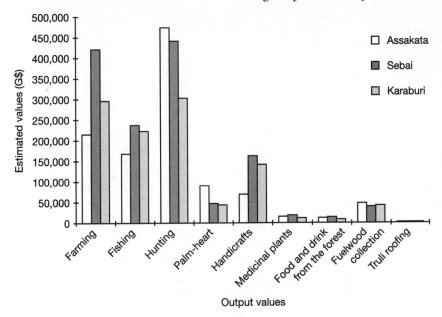

Figure 10.2 A comparison of the average household values of productive outputs in three Amerindian villages in North-West Guyana

Note: These household averages are based on participating households only.

Table 10.9 A comparison of bushmeat catches and household consumption in three Amerindian villages in North-West Guyana, 1996

Village	Total catches (lbs)	Average catches by participating households (lbs)	Total home consumption (lbs)	% of hunting catch used for home consumption	% of households participating in hunting	Average home consumption by participating households (lbs)	Average per capita consumption of bushmeat (lbs/annum)
Assakata	61,145	5,558	37,909	62	48	3,446	215
Sebai	77,386	3,364	32,502	42	74	1,413	162
Karaburi	28,411	1,353	11,080	39	24	528	20

Note:
Average home consumption rates are calculated on the basis of those households in the village which are involved in hunting. These figures represent food only, and do not include the value of trapping for the wildlife trade. Per capita figures are taken over the total population of the village.

Table 10.10 A comparison of fishing catches and household consumption in three Amerindian villages in North-West Guyana, 1996

Village	Total annual catch (lbs)	Average catch by participating households (lbs)	Total home consumption (lbs)	Average home consumption by participating households (lbs)	Average per capita home consumption of fish (lbs/annum)
Assakata	57,788	3,210	35,230	1,958	200
Sebai	64,444	2,222	34,155	1,178	170
Karaburi	74,433	2,567	50,520	1,742	90

Note:
Average home consumption rates are calculated on the basis of those households in the village involved in fishing, while per capita rates are taken across the total population of the village.

strong demand from the market at Port Kaituma, and the subsequently higher prices for meat and fish in that area. The low level of catches in Karaburi, and the low household consumption rates, are indicative of the relatively small proportion of households involved in hunting, and the fact that much of their output is sold both to other households in the village, and in a salted state to buyers in the market in Kumaka. This also perhaps reflects the fact that less wildlife is available in the Karaburi area since the population density in that village is much greater than in the others, and thus more pressure has been brought to bear on such resources in previous years, possibly leading to a greater degree of depletion.

It is interesting to note that in contrast to bushmeat catches, fishing catches in Karaburi are higher than in the other villages, possibly as a strategy to supplement the supply of dietary protein which is not available from other sources. On the basis of these figures, it can also clearly be seen that households in Assakata retain more bushmeat and fish for home consumption than do households in the other villages. This can possibly be explained by the fact that farming in that village is less productive than in the others, and also in Assakata, a higher proportion of farm output is consumed at home than in the other villages.

During the surveys carried out in these villages in 1996, householders, hunters and village elders were asked their opinions about their perceptions of the environment in which they live. From this, it was found that almost 90 per cent of hunters believe that hunting will be more difficult in future. When asked reasons for this difficulty, approximately 85 per cent felt that this was due to there being less animals in the future. This consciousness about animal and bird depletion rates was further elaborated on by the fact that 23 per cent of these hunters said this was likely to be the result of forest disturbance.

The value of forest inputs in alternative livelihood strategies

Those activities which are forest-dependent clearly make a contribution to the economy of the village, and by examining their monetary value, it is possible to assess the proportion of village output which depends on forest utilisation. From the figures for individual sectors shown in Table 10.11, the monetary value of plants collected from the forest for various uses is relatively low, but since these plants provide fuel, medicine, food and drink, roofing and handicraft materials, the full value of their worth must reflect all of these.[6] When added together, they take on a significant proportion of total output value, and by combining this plant value with the total estimated value of hunting and fishing, the total value of forest inputs reflects some measure of the importance of forest inputs as a factor of production in forest survival strategies.

The total value of forest inputs in each village has been derived through an input-output accounting process, and details of this procedure are shown in Box 10.1. Table 10.9 shows how forest inputs are used in each productive sector, and this shows how the value of forest inputs varies both between the activities involved, and between villages.

Taking all of the households included in this study, the figures indicate that the average net value of forest inputs in these study villages is G$322,744 (US$2343) per household, or G$49,308 (US$357) per capita. These amounts represent imputed accounting values which must not be confused with income flows. Such figures calculated in this way actually represent the value added by nature, to household labour and capital inputs. While these may seem insignificant values in global terms, their removal from forest households, through deforestation, does represent a very real threat to the possibility of ensuring the sustainability of forest livelihoods.

The interface with the global economy, and its impact on Amerindian livelihood strategies

'Market failure' is the term used by many economists to describe the mechanisms which perpetuate low standards of living in villages such as the ones described here. One of the reasons for the occurrence of this failure is said to be due to a failure of the institutional framework to facilitate the growth and redistribution of economic or political power. Communities such as these face many problems as a result of market and institutional failures, and these can be magnified by the effects of globalisation. As a result, these conditions may well prove to be the roots of a process of unwanted cultural harmonisation, irreversibly reducing the variability of possible livelihood opportunities for future human populations. In addition to this erosion of cultural diversity, these failures also have an eco-logical impact in the form of increased pollution levels, loss of biodiversity and unsustainable levels of consumption of renewable resources.

Few economies today do not use money as a medium of exchange, and these isolated Amerindian villages are no exception. Money acts as an interface with the

Table 10.11 A comparison of the net sectoral value of forest inputs (G$)

Village	Farming	Fuelwood collection	NTFP collection	Fishing	Handicrafts	Palm-hearts	Hunting	Total value of forest inputs (G$)
Assakata	871,442	165,038	169,247	1,016,657	-513,372	293,644	3,183,544	5,186,200
Sebai	7,129,100	187,356	625,457	4,159,035	310,989	-5,688	7,447,307	19,853,556
Karaburi	9,934,208	542,982	584,750	2,771,608	1,153,248	13,823	6,112,019	21,112,638

Note:
Negative values for handicraft in Assakata, and palm-heart in Sebai reflect the fact that in many households in those villages, labour hours are spent in these activities, but only on a small scale for household consumption, and thus no corresponding income is recorded for them resulting in a negative value here.

Box 10.1 The accounting framework

As a means of estimating the value of forest use, a model of the village economy is developed, and the value of the Gross Village Product (GVP) calculated. This is based on the usual accounting framework, as used in the United Nations System of National Accounts, but modified to represent the simpler economy found in a subsistence village. The model of the village economy is calculated on the basis of the usual equilibrium accounting assumption that:

Value of household inputs = value household outputs

Here:

$$household\ inputs = wL^h + rK^h + \delta K^h + p_f F^h$$

Where:

w = wage rate;
L^h = weighted hours worked by household h (weighted for men, women and child labour inputs);
r = rate of interest for the use of capital in production;
K^h = productive capital used by household h;
δ = capital depreciation rate;
p_f = implicit price of each unit of nature (forest) used;
F^h = implicit quantity of nature (forest) used by household h.

Similarly:

$$household\ outputs = \sum_{i=1}^{n} p_i Q_i^h$$

Here:

p_i = price of the good;
i = counter for NTFPs; hunting, fishing, handicraft and farming outputs, etc.;
Q_i^h = quantity of that good produced by household h.

All values used here refer to the period of one year, and so for convenience, the time subscript (t) usually applied will be omitted. The value of 'savings' (ΔK) would be included in this equation as an output, identified by one of the Q_i^h values, but without intertemporal household data, it is impossible to identify any specific value for capital accumulation by households. As a result, this value is included in the total of 'value added' associated with the use of the forest.

By equating the value of household inputs and outputs, we get:

$$w L^h + r K^h + \delta K^h + p_f F^h \equiv \sum_{i=1}^{n} p_1 Q_h^1 \qquad [1]$$

To build the complete model of the village, we then need to consolidate all the data together, and the Net Village Product (NVP), is obtained by summing across all households h:

$$NVP = \sum_{h=1}^{H} (w L^h + r K^h + \delta K^h + p_f F^h) = \sum_{h=1}^{H} \sum_{i=1}^{n} p_i Q_i^h \qquad [2]$$

Note: The Gross Village Product is converted to the Net Village Product through the process of depreciation, included in Equation [1] as δK^h.

Determining the value of forest inputs from Net Village Product
The value of $p_f F^h$ will be derived as a residual from the completed equation of all other inputs and outputs. This residual represents the contribution made to NVP by the various forest products, and each of these are in the form of output values from village activities generated by the use of forest resources. This is shown as equation:

$$p_f F^h = \sum_{h=1}^{H} \left(\sum_{i=1}^{n} p_i Q_i^h - (W L^h + r K^h + \delta K^h) \right) \qquad [3]$$

global economy, and the lack of it has a significant influence on the lives of people in Amerindian villages. It is needed to buy items such as soap and matches, as well as clothing, shoes and writing materials for children to go to school. Although such purchases may represent a small cash requirement, they can nevertheless have the effect of putting economic pressure on households in general, and women in particular, as they are often the ones to be more involved with domestic arrangements and schooling. As a result, households may be driven to unsustainable behaviour in terms of wildlife and fuelwood extraction, and inappropriate farming practices. In order then to achieve some meaningful degree of sustainability in such communities, these processes must be first understood, and the market and institutional failures identified and corrected.

Levels of household capital and its distribution in each village

Details of capital holdings were collected from 100 per cent of households in each village, and these were assessed on the basis of ownership of important productive items. The total value of productive capital in each village is shown in Table 10.12, and this reflects the productive capacity of the villages, as indicated by the correlation coefficient of 0.98 between total productive capital and the value of village output.

Table 10.12 Capital availability, 1996 (G$)

	Assakata	Sebai	Karaburi
Total village productive capital	570,374	1,401,300	2,880,204
Average productive capital, by household	21,937	45,203	32,362
Cost of total village productive capital use p.a.	159,705	392,364	806,457

As in any other economic system, those using productive capital most efficiently will accumulate more wealth, and this can be seen to have occurred in a number of households in these villages. Although total productive capital is clearly greatest in Karaburi, where there are many more households than in the other villages, the level of productive capital available to households on average, is greatest in Sebai. Although this correlates with the fact that Sebai is the most productive of the three villages, production is very labour intensive in all of them, so the variation in levels of capital use has only marginal effects on output relative to the role played by labour.

The distribution of capital stock

How productive capital is distributed between the households in the three study villages is shown in Figure 10.3, demonstrating the approximately lognormal distribution which illustrates the spread of wealth within many economies (Lambert 1993). This indicates that the majority of households have low to middle wealth holdings, with just a few households reaching higher levels. In all cases, those few households with greater productive capital stock are those owning valuable equipment such as chain saws or outboard engines.

Capital is an essential factor of production, and lack of it retards economic growth. The low cash flows in these forest economies mean that it is very difficult for such groups to mobilise capital for productive use, or as a cushion in times of need. In economic terms, this suggests the occurrence of a failure in the capital market, and it results in inability for even successful households to break away from subsistence. This is a problem which possibly could be overcome by bringing about the institutional changes necessary to promote the development of micro-credit and saving systems, already becoming widespread in many regions of the world.

New livelihood opportunities

For communities living in forests like those in Guyana, a number of possible future livelihood opportunities exist. These include palm-heart extraction, bush-meat production, fish farming for food and collector species, orchid and exotic plant cultivation and export, cultivation and packaging of medicinal plants, eco-tourism, and production of organic pesticides. While some of these may require significant research and development, the current operation of the palm-heart

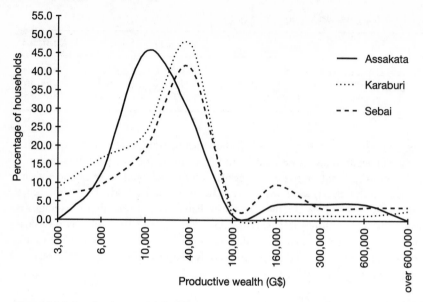

Figure 10.3 Productive wealth holdings in three Amerindian villages in Guyana, 1996

industry in Guyana provides one example of how these alternative development options operate in practice.

Palm-heart extraction

Today, the extraction of palm-heart from the manicole palm (*Eurtepe oleracea*) is of growing importance as a forest activity in the North-West of Guyana. This began in 1987, with the granting of an extraction concession of some 50,183 hectares covering the Barima, Waini, and Baramanni rivers, and part of the Kaituma and Aruka rivers. This concession was taken up by the Amazon Caribbean Company (known locally as AMCAR), which constructed a cannery on the Barima river, and began commercial production in 1989. Levels of palm-heart harvesting in Guyana are shown in Table 10.13.

The royalty payment from palm-heart harvesting amounted to 1.1 per cent of all royalty payments received by the Government in 1995. Since these include all payments for all types of timber extraction, and other forms of royalties (Bank of Guyana 1995), it seems likely that environmental damage arising per dollar of government revenue is likely to be less for this industry than for other extractive processes. Nevertheless, some concern has been expressed about the rapidly increasing rate of manicole harvesting (Forte 1996b), and this has led to extensive biological and ecological research[7] being conducted on the manicole plant. Although the results of this study are not yet published, preliminary evidence (Johnson 1994) suggests that this form of forest exploitation is much more sustainable than conventional timber harvesting, simply on the basis of palm

Table 10.13 Palm-heart harvesting in Guyana, 1990–95

Year	Harvest level (stems)	Royalty payment (G$)
1990	162,679	40,669
1991	152,750	55,625
1992	284,789	284,789
1993	2,715,677	2,715,677
1994	5,946,633	5,946,633
1995	6,190,456	6,190,456
Total	15,452,984	15,233,849

Source: Figures provided by A. A. Rojan, Statistical Officer, on authorisation from the Commissioner of Forests, Guyana Forestry Commission, 15 August 1996.

Note:
Royalty rates increased from G$0.25 per stem in 1990 to G$1 per stem in October 1991.

regeneration rates. It seems likely that only a 5-year period of regrowth is needed between harvests, and as long as harvesting is conducted in a responsible manner, the harvesting of palms is possibly one way in which forest resources can be used sustainably in the North-West of Guyana.

Royalty payments are made by stem to the Guyana Forestry Commission, and although some dispute with the company has arisen over the definition of the term 'stem' (and consequently the royalty payments), production has risen considerably since that time. By mid-1996, this concession had produced a total royalty payment for the Government of G$15,233,849 (GFC 1996). In addition to these revenues earned by the Government through the collection of royalties on palm-heart, Guyana also benefits from this industry in the form of employment opportunities and workers' wages, as well as the foreign currency earnings from the export of canned palm-heart. According to company records,[8] in July 1996, a total of 184 workers were employed in the canning factory in Drum Hill, on the Barima River, with average basic wages ranging from G$6,180 per month to G$31,000 per month (plus bonuses), depending on the type of work done. All workers receive free food and lodging while at the plant, and while on leave they receive a leave bonus of 10 per cent of what they have earned in the previous period. Over 1,000 cutters supply the palm-hearts to the company, and their harvests are purchased by agents and transported by boat to the plant three times per week. The cutters all work on a part-time, freelance basis, which according to company management is to ensure that the work does not interfere with the normal farm work of their households. Of the cutters interviewed in the three villages of this study, it was felt that the company made an important contribution to their lives, and they did not want the business to cease, since it provided them with a rare opportunity to earn a cash income.

From the macroeconomic viewpoint, the industry also has a positive multiplier effect, with all contracted employees paying income tax at standard rates, and export duties being charged on all cans leaving Guyana. In July 1996, a total of 48,716 large cans (520 g) were produced in the Drum Hill plant, along with 127,998 small cans (220 g). Additional output is produced in jars at a smaller operation on the East Bank of the Demerara river, and at the new canning plant in Berbice.

Some observers have suggested that the palm-heart industry is destroying trad-itional ways of life, as factory workers live on-site for extended periods, possibly neglecting their domestic responsibilities back in the villages. While it does seem likely that this may apply to some individuals, the numbers involved are small, and thus unlikely to have a widespread effect. It is inevitable that new ways of life may bring about changes in norms of acceptable behaviour in communities, but it seems likely that cultural breakdown resulting from the activities of palm-heart extraction will be less damaging than that caused by large-scale logging oper-ations. In both cases however, centralised places of employment are often the source of sexually transmitted diseases which get carried back to the villages and weaken the home populations.

The potential of biodiversity

Guyana is a member of the International Tropical Timber Organisation, and as a signatory to both the ITTO agreement (ITTO 1992) and the Convention on Biodiversity, it is committed to forming a National Biodiversity strategy, requir-ing *in-situ* conservation of key species. This is taking place to some degree within the Iwokrama Forest Research Programme, but questions about stakeholder involvement and intellectual property rights are yet to be resolved. With support from the Global Environmental Facility, it is likely that at least some progress towards such biodiversity conservation will be achieved, and if some income-generating potential of this can be released, livelihood opportunities for Amerindian communities will be increased.

The economic importance of biodiversity has been much discussed (Brown *et al.* 1993; Pearce and Moran 1994), and it seems clear that this is of crucial importance to future human development. The ability of Amerindian populations to capitalise on their indigenous knowledge of animal and plant use for resins, fibres and medicinal purposes, depends both on their own entrepreneurial skill, and the development of appropriate institutional mechanisms. Examples of how this knowledge can be applied could include the development of organic fungicides or pesticides, or to domesticate indigenous wild animals for bush-meat production.

Community views about the future

Generally, there is an *a priori* assumption that people who live in forest environ-ments will feel some affinity with the forest as an entity, resulting in some implicit measure of value placed upon it. While it is impossible to estimate this in the conventional economic sense, it is important to try to understand more about it. One way of doing this is to examine how such forest dwelling people perceive their environment, and how they feel about life within it. It is with this in mind that questions on such issues were included in this case study, and it is interesting to note that, in spite of the subsistence lifestyle that they live, the majority of the people in these villages consider themselves as 'happy', although a higher proportion of women say they are happy than do men. Some quite significant variation exists in people's attitudes towards the future of life in the village, as indicated by their responses to questions concerning their views about their children. In general, men seem to want the children to stay in the village more than women do, with 73 per cent of men wanting this, although when asked about prospects for the future, over 60 per cent of both men and women suggested that the future is likely to be easier than the present.

How people think things in the village should change

When asked what they thought they would most like to change in the village, responses were quite variable. For men, most improvement in life in the village would arise from an improvement in the standard of living, better transport and better health. In addition to these changes, more women than men would also like to see more shops or businesses, as well as other lifestyles, and also more women would like nothing to change in the village. This last point implies that the maintenance of a traditional lifestyle is perhaps more important to some women than to men. It is interesting to note that to improve their own lives, the issues of better health, education and extra money are mentioned more frequently by men than by women, while more women than men suggest such improvements as a better home, better health, and getting a job. These attitudes may influence what people would like to see happen in the process of development, and so a number of development options have been examined in more detail in the next section.

Peoples' perceptions of development options in the villages

The process of development in villages such as these is often met with mixed feelings, and in the past, local inhabitants have rarely been consulted. Some people in such villages see 'development' as the opportunity to improve their own standard of living, while others associate it with the exploitation of natural resources, without benefits for local people. In order to assess how people in the villages of this study view alternative development priorities, views were sought on a number of possible options.

In Table 10.14, men's and women's views on various development issues in the villages are shown. The numbers shown here indicate the mean scores assigned to each issue by all respondents (representing 143 households), and their standard deviations, and suggest that both men and women see some variation in how such issues influence their own families and the community. Indications of how they

Table 10.14 The relative importance of development issues, as indicated by mean
scores assigned by men and women (standard deviations shown in parentheses)

Important for:	Education	Health	Nature and environment	Business development	Agricultural development	Tourism
Men's views						
The family	4.70	4.92	4.35	4.25	4.83	2.67
	(0.80)	(0.37)	(0.81)	(0.97)	(0.48)	(1.90)
The commmunity	4.95	4.93	4.57	4.30	4.94	3.10
	(0.22)	(0.35)	(0.89)	(0.80)	(0.23)	(2.05)
Children's	4.94	4.95	4.71	4.50	4.81	3.46
lifetime	(0.49)	(0.22)	(0.53)	(0.69)	(0.49)	(1.97)
Women's views						
The family	4.76	4.93	4.31	4.31	4.93	3.01
	(0.87)	(0.35)	(1.12)	(0.93)	(0.37)	(1.97)
The commmunity	4.92	4.92	4.52	4.45	5.00	3.41
	(0.50)	(0.37)	(1.07)	(0.87)	(0.00)	(2.06)
Children's	4.86	4.94	4.34	4.28	4.87	3.17
lifetime	(0.76)	(0.33)	(1.04)	(0.93)	(0.43)	(1.97)

Note:
Figures show the mean scores assigned to each of the issues shown, where respondents gave a
score from 0 to 5 to indicate the level of importance of each issue.

may view future developments are given by the scores relating to the importance
for these issues in their children's lifetime.

From these figures, it can be seen that health and education are seen as most
important, for both men and women, with agricultural development also rated
quite highly. Responses relating to business development and tourism are much
less clear cut, with greater variability indicated by the larger standard deviations
on those questions. If tourism is to become important in the future, this will bring
about some new employment opportunities for forest dwellers, but if village
development is to represent the residents' views, it is important that these variations should be further examined before such changes are introduced.

The importance of carbon sequestration

The process of carbon sequestration (the absorption of carbon by vegetation), is a
basic element of atmospheric maintenance, and thus is an important environmental service provided by forests. In addition, carbon is locked in trees and other
vegetation, so deforestation contributes to global warming by releasing this carbon into the atmosphere. Estimates of the damage done by global warming have
been made (Frankhauser 1995), and some studies have attempted to assess the
value of 'carbon credits' in standing forests (Frankhauser and Pearce 1994). As a
result of this work, it can be concluded (Pearce 1996) that the largest degree of
anthropocentric use-value of tropical forests is that attributable to carbon storage

and, in a country such as Guyana, where so much land space is covered by forests, this has important potential for carbon trading, given current expectations on global warming (IPCC 1996). If and when these institutional changes are implemented, further possible livelihood opportunities may come about for these communities, involving some kind of employment as 'forest wardens' to monitor forest use.

Conclusions

In traditional societies throughout the world, important beliefs and cultural heritage, often with deeply rooted ecological significance, are passed on from one generation to another in the form of myths and legends. Studies from a number of African states have recorded these, and it is well known that forests have important spiritual, religious, political and social values for a number of West African tribes (Falcolner and Koppell 1990). Extensive literature exists on cultural values and beliefs from Central and Latin America, and today interest in this is often concerned with questions of intellectual property rights of indigenous forest people. Examples of this are found in the extensive writings of anthropologists such as Darrell Posey (Posey 1985; Posey *et al.* 1995), while the importance of this knowledge for biodiversity conservation has been highlighted by ecologists such as Gadgil and Folke (Gadgil *et al.* 1993).

More effective utilization of the indigenous knowledge of these communities holds great potential for the future of these people, as long as the forest resources themselves continue to survive, and the institutional frameworks of the developed world are extended to provide mechanisms by which this potential can be realised. Evidence that institutional change is needed to facilitate this, is provided by publicity in the USA (Reuters 1998) about attempts by individual American citizens to take out patents on indigenous medicines, as developed by Amerindians. According to COICA (Coordinating Secretariat of Organisations for Indigenous People in the Amazon), the Inter-American Foundation (an agency set up to help indigenous peoples and funded by US tax payers) was in support of this violation of their intellectual property rights. This concerned the plant Banisteriopsis Caapi, also known as Ayahuasca or Yagi, widely used by numerous Amerindian groups throughout Central and South America (COICA 1998), and now known to have important medicinal potential.

In addition to institutional changes to protect intellectual property rights, general institutional strengthening could address poverty issues through the development of more effective and reliable trading networks, and assistance in the development of mechanisms to promote the adoption of new production strategies. An attempt to identify the factors which have enabled this lifestyle to be so enduring should be made, in order to ensure that the prerequisites for it are not lost through resource mis-management today (Turner *et al.* 1998). To support this, research can be promoted to assess the potential for a 'sustainable loop' system of slash-and-burn agriculture, and the potential development of plant-based pesticides and fungicides.

It is clear that for forest dwellers throughout the tropics, livelihoods depend heavily on the utilisation of forest products and services, just as they do in these Amerindian communities. In the face of the continuous pressures being brought on them as a result of both small- and large-scale enterprises, the traditional way of life of Amerindians in Guyana is seriously under threat today. If the full meaning of sustainability is really to be achieved, it is important that the range of their environmental entitlements is secured (Scoones 1998), and that the necessary institutional arrangements be made to consolidate this for the future (Binswanger 1998). Given the stability and possible sustainability of these traditional liveli-hoods, and the expertise about the forests which they embody, it is foolish for this important social group to be neglected, and while poverty alleviation is an important first step to be addressed if further weakening of this way of life is to be avoided, it is essential that institutional changes be made to overcome the developmental problems which they face.

Notes

1   Including those lying beneath titled Amerindian areas (as outlined in Section 20A of the 1976 Amerindian Act).
2   This assertion is made on the assumption that current costing procedures and fiscal policies will inherently be flawed, due to a failure to take full account of all environmental externalities in their calculation.
3   Clearly data collected in a snap-shot approach requires aggregation, and through-out this work, every attempt has been made to take account of seasonal variations in this procedure.
4   Prices taken in July 1996, at which time the exchange rate for the Guyana dollar was G$136 = US$1, or G$208 = UK£1.
5   Labour values are calculated on the basis of the 'shadow price' of labour, which in this case is determined by the amount which can be earned by any worker as a result of freelance palm-cutting. In 1996, this amounted to G$70 per man hour. Household labour values are determined by the individual demographic com-position of each household.
6   In addition to these monetary values, another factor to be considered is that the monetary valuations of forest foods and drinks included here do not reflect their full value, as no account is taken of their importance as an essential supply of vitamins and minerals to the often impoverished diet associated with subsistence households.
7   The work of Tinde van Andel has been part of the Tropenbos-Guyana Pro-gramme, funded by the Tropenbos Foundation of the Netherlands.
8   Information provided by Samuel Arjune, Assistant Manager, AMCAR, Drum Hill, in an interview with the author, in July 1996.

References

Bank of Guyana (1995) *Statistical Bulletin*. Bank of Guyana, Georgetown.
Binswanger, H.C., (1998) Making sustainability work. *Journal of Ecological Econom-ics*, 27: 3–11.
Brown, K., Pearce, D.W., Perrings, C. and Swanson, T. (1993) *Economics and the Conservation of Global Biological Diversity*. Working paper no. 2. Global Environ-mental Facility, Washington, DC.

Bruhns, K.O. (1994) *Ancient South America*. Cambridge University Press, Cambridge.

Carney, D. (ed.) (1998) *Sustainable Rural Livelihoods: What Contribution Can We Make?* DfID, London.

COICA (1998) (Coordinating Secretariat of Organisations of Indigenous Peoples from the Amazon) *The Patent on the Sacred Plant Ayahuasca*. Open letter to the President of the Inter-American Foundation, Washington. Published on the Internet site http://forests.lic.wisc.edu/gopher/guysurin/sacplant.txt.

Colchester, M. (1993) *Who's Who in Guyana's Forests*. World Rainforest Movement, Oxford.

Daly, H. (1989) Towards a measure of sustainable social net product. In: Ahmad, Y.J., El Serafy, S. and Lutz, E. (eds), *Environmental Accounting for Sustainable Development*. World Bank, Washington, DC.

DeFilipps, R.A. (1992) The history of nontimber forest products from the Guianas. In: Plotkin, M. and Famolare, L. (eds), *Sustainable Harvest and Marketing of Rain Forest Products*. Island Press, Washington, DC.

Falconer, J. and Koppell, C. (1990) *The Major Significance of 'Minor' Forest Products*. FAO, Rome.

Fanshawe, D.B. (1948) *Forest Products of British Guyana, Part II, Minor Forest Products*. Forestry Bulletin, no. 2, Forest Department, Georgetown.

FAO (1989) *Forestry and Nutrition, A Reference Manual*. FAO, Rome.

Flaming, L., Cassells, D. and Godoy R. (1995) *An Economic Analysis of the Timber Industry in Guyana*. Harvard University, Harvard.

Forte, J. (1995) (ed.) *Indigenous Use of the Forest – Situation Analysis with Emphasis on Region 1*. British Development Division in the Caribbean, Barbados, and University of Guyana Amerindian Research Unit, Georgetown.

Forte, J. (1996a) *Thinking about Amerindians*. Private publication, Georgetown.

Forte, J. (1996b) *About Guyanese Amerindians*. Private publication, Georgetown.

Frankhauser S. and Pearce, D.W. (1994) The Social Costs of Greenhouse Gas Emissions. In: OECD, *The Economics of Climate Change*, OECD, Paris, 71–86.

Frankhauser, S. (1995) *Valuing Climate Change*. Earthscan, London.

Gadgil, M., Berkes F. and Folke, C. (1993) Indigenous knowledge for biodiversity conservation. *Ambio* 22, 151–56.

Gaia Forest Conservation Archive (1998) URL http://forests.org.

GFC, (1996) *Palm-heart Harvesting and Royalties*. A.A. Rojan, Statistical Officer, Guyana Forestry Commission, 15.8.96.

GOG (Government of Guyana) (1989) Mining Act, Georgetown.

GOG (Government of Guyana) (1993) *Household Income and Expenditure Survey*. Government of Guyana Statistical Bureau, Georgetown.

GOG (Government of Guyana) (1995) *Statistical Bulletin*, (4) 4. Bureau of Statistics Georgetown.

Hill, P. (1986) *Development Economics on Trial – The Anthropological Case for a Prosecution*. Cambridge University Press, Cambridge.

Hills, T.L. (1961) *The Interior of British Guiana and the Myth of El Dorado*. Canadian Geographer, 5: 30–43.

IPCC (Intergovernmental Panel on Climate Change) (1996) *Second Assessment Report*, Vol. 1. Cambridge University Press, Cambridge.

ITTO (1992) *Criteria for the Measurement of Sustainable Tropical Forest Management*. Policy Development Series No. 3. ITTO, Yokahama.

Johnson, D.V. (1994) *Report on the Palm Cabbage Industry in North West Guyana.* Palm Specialist Group, IUCN, Silver Spring, USA.

Lambert, P.J. (1993) *The Distribution and Redistribution of Income, a Mathematical Analysis.* Manchester University Press, Manchester.

Markandya, A. and Perrings, C. (1993) *Accounting for an Ecologically Sustainable Development: A Summary.* UNEP, Environmental Economics Series paper no. 8.

Melnyk, M. (1995) *The Contributions of Forest Foods to The Livelihoods of the Huottuja (Piaroa) People of Southern Venezuela.* PhD. Thesis, ICCET, University of London.

Mendelsohn, R. and Balik, M. (1995) The value of undiscovered pharmaceuticals in tropical forests. *Economic Botany* 49 (2), 223–28.

Pearce, D. (1996) *Can Non-Market Values Save the World's forests?* Paper presented to the Symposium on Non-Market Benefits of Forestry, UK Forestry Commission, Edinburgh, June 1996.

Pearce, D. (1998) Private communication with the author, May 1998.

Pearce, D. and Moran, D. (1994) *The Economic Value of Biodiversity.* Earthscan, London.

Pearce, D.W. and Turner, R.K. (1990) *Economics of Natural Resources and the Environment.* Harvester-Wheatsheaf, London.

Pons, T. (1996) Netherlands Coordinator for the Tropenbos Guyana Project. Private communication with the author.

Posey, D. (1985) Indigenous management of tropical forest ecosystems: the case of the Kayapo Indians of the Brazilian Amazon. *Agroforestry Systems* 3, 139–58.

Posey, D., Dutfield, G., and Plenderleith, K. (1995) Collaborative research and intellectual property rights. *Biodiversity and Conservation* 4, 8.

Reuters News Agency (1998) *US Agency Criticised over Amazon Plant Patent.* Washington. 4 March 1998.

Scoones, I. (1998) Sustainable Rural Livelihoods: A Framework for Analysis. IDS Working Paper no. 72 IDS, Brighton.

Sullivan, C.A. (1997) *The Economic and Social Value of Non-Timber Forest Products in a Forest Village Economy in N.W. Guyana.* Tropenbos-Guyana Reports, 97–4. Utrecht.

Turner, R.K., Adger, W.N. and Brouwer, R. (1988) Ecosystem services value, research needs, and policy relevance: a commentary. *Ecological Economics* 25, 61–65.

UNCED (1992) *Report of the United Nations Conference on Environment and Development.* United Nations, New York.

World Commission on Environment and Development (1987). *Our Common Future,* Oxford University Press, Oxford.

WRM (1997) *Undermining Indigenous People and the Environment in Guyana.* World Rainforest Movement, Montevideo, Uruguay.

Index

Page numbers in italics refer to Figures and Tables

Adams, J. 24
Agenda 21 76, 86, 87, 158; *European Sustainable Cities* 70; and Healthy Cities project 83
agriculture: education 151–2; England and Wales 96–7, 98, *99*, 100–5; Guyana 160, 163–4, *166*, *167*, *168*; intensification 58; Lesotho 155; Nigeria 128–9, *129*; and perceptions of nature 23; Upper Canada 110–11, 113–14
Alberti, M. 60
Amerindians 158–60, 185–6; economic encroachment 160–2; lifestyles case study 163–79; livelihood opportunities 179–85
animal rights 26
anthropocentric ethics 25–6, 28, 32
Association for Public Health 88

Banisteriopsis Caapi 185
Barnett, H. 39, 40
Barrow, M. 38
Beck, U. 24
Beinart, W. 107
Bennett, J.W. 119
Bigsby, John 116
biocentric ethics 25, 26, 27, 28, 32
biodiversity: Guyana 182; loss 21–2; maintenance 47–8; World Biosphere Reserves 49
Bishop, R. 46
Breheny, M. 60
Broad, R. 53
Broken Hill Proprietary 162
Brundtland Commission 2

Caird, J. 95
Canada 108, *see also* Upper Canada
capital 178–9, *179*, *180*
carbon sequestration 184–5
Careless, J.M.S. 110–11
carrying capacity 47, 61–2
Catton, W. 61
CFCs 21
Chambers, Robert 53
Chlorocardium rodiei 161
choices 8, 11
Ciriacy-Wantrup, S.V. 45–6
cities *see* urban environment
city health profiles 83
CLAIRE initiative 71
climate change 20, 47, 130, 184–5
Coase, Ronald 41
Coates, P. 107
Cobb, J. 52
Codd, Francis 107, 117–19
Coke, Lieutenant 116
Cole Harris, R. 111, 112
community organisation 152–3
conservation 7–8, 47–9
consumption 18, *19*, 52; environmental effects 5–7, 29; and happiness 22; health risks 77
Contaminated Land – Applications in Real Environments 71
contingent valuation methods 43
Convention on Biodiversity 182
Convention on International Trade in Endangered Species 49
Council for the Protection of Rural England 63
Cowan, Helen 109, 110
Craig, Gerald 113, 114

critical natural capital 3
critical zone concept 45
Cullingworth, J.B. 68
cyanide 161–2

Daly, H.E. 47, 52
Dasgupta, P. 38
Daubeny, Professor Charles 116
democracy 29, 30–1, 32
Department of the Environment 60, 86;
    Habitat II report 70; *The Indicators of
    Sustainable Development for the United
    Kingdom* 82; Planning Policy
    Guidance Notes 69; *Sustainable
    Development* 68–9; *This Common
    Inheritance* 68
Department of Environment, Transport
    and the Regions 71
Department of Health 86
development 29, 52; Amerindians 183–4
Diamond, P.A. 43
discounting 52
displaced consumption 6
Dunlap, R. 7

eco-warriors 29
ecocentric ethics 25, 26–7, 28, 30, 32;
    and environmental law 28–9
ecological analysis 81
ecological footprints 6, 61–2
ecological law 28–9, 32; and democratic
    theory 30–1; and education 31–2; and
    liberal theory 29–30
ecological modernisation 7, 18
economic development: and
    environmental sustainability 53–4;
    health impacts 75–6; and health
    inequalities 78–9, 81
economic efficiency: and environmental
    benefits 8
economic encroachment: Amerindians
    160–2
economic sustainability 45; Nigeria
    138–9
economics, and the environment 36–44
ecosystems 26–7, 40; protection 47–8,
    see also ecocentric ethics
education: and ecological law 31–2;
    Nigeria 132, 134–5, *135*, 137;
    Southern Africa 144, 145, 147–56
Elkin, T. 60
Elliot, J.A. 150
emigration 106, 108–9
emissions 48, 49

energy 29, 52
entitlements 11
environment 4–5; spheres of activity 5–7,
    77–8
environmental accounting 42–4
environmental activists 29
environmental damage 17, 18, 35, 40–1,
    48; costs 42–4; Nigeria 124
environmental ethics 24–5, 28, 32, 46;
    anthropocentric 25–6; biocentric 25,
    26, 27; and democracy 31; ecocentric
    25, 26–7; education 31
environmental functions 44–5
environmental health 75–6, 80–2, 88–9;
    Healthy Cities project 82–4; policy
    frameworks 85–8; and sustainability
    84–5
environmental knowledge 19, 20–2
environmental law: and ethics 28–9; and
    liberalism 29, *see also* ecological law
environmental policy 7–8, 47–9
environmental resources *see* resources
environmental risk 24, 46
environmental sustainability: Nigeria
    126–30; principles and standards
    44–9; and sustainable development
    51–5
environmental values 43
ethical sustainability 131; Nigeria 133–7
ethics *see* environmental ethics
European Commission 50, 53, 68
European Community Expert Group on
    the Urban Environment 70
European System of Integrated
    Economic and Environmental Indices
    50
Evershed, Henry 100
exchange value 51–2
existence values 43
experts: knowledge 8; risk 24
externalities 41, 52
extinction 40

farming *see* agriculture
Fidler, Isaac 113, 115–16
Folke 47
food: England and Wales 95, 97, 104–5;
    Guyana 159–60
food miles 6
forests: Guyana 158–61; in livelihood
    strategies 175, *176*, *177–8*; in
    production process 169, 171, *172*;
    Upper Canada 110, 112; valuations of
    resources 163

Fowler, Thomas 115
Freyfogle, E.T. 30
Friends of the Earth 127–8
frontier society, Upper Canada 107, 115–19

Gaia 23
Gambari, I.A. 124
General Agreement on Tariffs and Trade (GATT) 53
German Advisory Council on Global Change 49
global warming 20, 184–5
gold 161
Golden Star Resources 162
Goldman, A. 128–9
Greenfield Campaign 63
Greenheart 161
greenhouse gases 20, 58
Guyana 158–60, *165*, 185–6; lifestyles case study 163–75; livelihood opportunities 179–85

Ha Rateme 144–5, 147
Habermas, J. 4
Habitat II 70
Hall, D. 39–40
Hall, J. 39–40
Hancock, T. 84–5
happiness 22
Hausaland 123, 125, *126*, 139–40; economic sustainability 138–9; environmental sustainability 128–30; social and ethical sustainability 131–7
Hausman, J.A. 43
Heal, G. 38
Healey, P. 68
health 75–6; geographical variations 78–80, *see also* environmental health
Health for All 2000 86, 87
Health for All Network 88
Healthy Cities project 82–4
Hirsch, F. 7, 22
Holdren, J.P. 60
Hotelling, H. 37, 38
Howison, John 113–14, 115
Hueting, R. 49
Human Development Index 52

Index of Sustainable Economic Welfare 52
indicators: environmental health 82; sustainability 4, 49–51
individualism 30

infrastructure: environment activities 6, 7; health impacts 77–8
institutional change 53, 185
intellectual property rights 185
intergenerational equity 8, 35, 54
International Monetary Fund 53
International Tropical Timber Organisation 182
International Union for the Conservation of Nature and Natural Resources (IUCN) 31
intragenerational equity 8, 35
intrinsic value 27

Jacobs, Jane 65
Jevons, W.S. 39

knowledge 19–32
Knowles, V. 108–9
Kohl, S.B. 119
kulle 125, 134–6, 137

labour migration, Southern Africa 145–7, 155
labour supply, Guyana 164–5, 169, *170*
land 59
land ethic 27
land policy: Southern Africa 146; Upper Canada 109–10, 111–12
land use 9, 58; mixed 65; Staffordshire 98, *99*, 100, *100*; sustainable 60–1, *see also* planning
landscapes, preservation 48
laws 18–19, *see also* environmental law
Leopold, A. 27
Lesotho: agriculture 155; education 145, 148, 152; employment 153; land availability 146, *see also* Ha Rateme
liberalism 29; and ecological law 29–30, 32
life chances 5; Nigeria 123, 133, 137; Southern Africa 147–50, 154; Upper Canada 116–19
lifestyles 5; choice 22–3, 32; and education 153–4, 155, 156; and risk 24, 32; and sustainability 144
livelihoods 5, 150; Amerindians 158; and lifestyle 153; Nigeria 123, 137, 139; Southern Africa 150–1, 155, 156; sustainable 10, 11, 12
Local Agenda 21 87
Local Government Management Board 88

Macdonald, N. 109
Malthus, Thomas 36–7, 39
market choices 8, 11
market failure 175
materials, circular flows 52
Meadows, D.H. 40
Mill, John Stuart 40–1
mining 161–2
modernism 1–2
money 175, 178
Montreal Protocol 49
morality 18
Morse, C. 39, 40
Moss, G. 59, 67
Movement for the Survival of the Ogoni People (MOSOP) 127
Mzunga 144–5, 147

NAMEA (National Accounting Matrix including Environmental Accounts) system 50
National Environmental Health Plan 88
natural capital 3
natural law 30
nature 6, 7; anthropocentric view 25; health impacts 78; perceptions 23–4; socialisation 1, 107; Upper Canada discourses 112–14
needs 2, 22–3
Nigeria 123; economic sustainability 138, *138*; environmental sustainability 126–8; social and ethical sustainability 131, 133; sustainability 124–5, *see also* Hausaland
Nigerian Environment Study Team (NEST) 128
non-renewable resources 47, 48
North America 10–11, 106–7, *see also* Canada

oil 127
Omai river 161–2
option values 43
Our Common Future 158
Our Healthier Nation 86–7, 88
ozone layer: depletion 21, 49; safeguarding 47

palm-heart extraction 180–2
Pearce, D. 39, 43
physical sustainability 6, 7, 78
Pigou, A.C. 41
Pinhey, Hamnett Kirkes 109, 118

planning 58, 71–2; policy responses 67–71
Planning Policy Guidance Notes (PPGs) 69
population 36, 37, 58; England and Wales 95, 97
poverty 53, 54
power 7–8
Precautionary Principle 49
production: conflicts 11; and education 152; environmental effects 5, 6; health impacts 77, 85
productive capital 178–9, *179*, *180*
Public Health Trust 88

quality of life 75

Rees, W.E. 61
Regan, T. 26
renewable resources 38–9, 47, 48
reproduction 29, 151
resilience 47
resources: accounting and valuation 42–4, 163, *177–8*; exhaustibility 37–40; property rights 41, 185; safe minimum standard 45–6, *see also* non-renewable resources; renewable resources
Ricardo, David 36, 39
Rio Summit 53, 88
risk 24, 46
risk analysis 81–2
Rolston, H. 27
rural communities 144, 145, 155

safe minimum standard (SMS) 45–6, 48–9
Sagoff, M. 29
scarcity 4
scientific knowledge 20–2
sectoral politics 8
SGAP$_P$ 49–51
Simcoe, John Graves 108
Simon, D. 155
Smith, S. 128–9
Smith, V. 38
Snow, Dr John 80
social capital: environmental activities 6, 7; health impacts 77–8
social sustainability 54, 131; Nigeria 131–2
society 8; economisation 1; naturalisation 107
Southern Africa 145–7

Soyinka, Wole 124
spatial variability, health 78–80, 81
species egalitarianism 26, 35, 36
species loss 21–2, 40
Staffordshire: agriculture 98, *99*, 100–5;
    food market 97; price indices 97–8, *97*
stewardship 25
Sulphur Protocol, Second 49
Susskind, L. 60
sustainability 44, 60; environmental
    health 84–9; Nigeria 123, 124–5,
    139–40; normative aspects 24–5;
    planning objectives 68–9, 70; positive
    aspects 24; rural communities 144,
    155–6; temporal dimension 9–10, *see
    also* economic sustainability;
    environmental sustainability; physical
    sustainability; social sustainability
Sustainability Gap 49–51
sustainability indicators 4
sustainable development 1, 2–4, 51–4,
    60
*Sustainable Development: The UK
    Strategy* 87
Sweden 50

Talbot, Thomas 108
Tansley, Arthur 27
Taylor, P.W. 26
technology: and environmental damage
    18, 48; progress 36–7, 47
Theroux, Paul 124–5
time series analysis 81
Times, The 63
Total Economic Value 43, 46
traffic 84, 85
transport, in cities 66–7
Trollope, Anthony 115
Turner, R.K. 39, 47

understanding: ethical systems and
    sustainability 19, 24–8; human
    perceptions and motivations 19, 22–4;
    incentives and restrictions 19, 28–32;
    natural world 19, 20–2
United Nations: Agenda 21 70, 76, 83,
    86, 87, 158; Handbook on Integrated
    Environmental and Economic
    Accounting 42

United Nations Conference on
    Environment and Development 49
United Nations Development
    Programme: Human Development
    Index 52
unsustainability 123
Upper Canada 11, 107, 108–11; frontier
    society 115–19; nature 112–14;
    settlement and the market 111–12
urban environment 59; carrying capacity
    61–2; derelict and damaged land 67;
    expansion 62–3, *64*; form 63, 65–6;
    health risks 77–8; sustainability 60;
    transport 66–7
use values 43
utility 22, 38

values 43
Van De Veer, D. 26

Wackernagel, M. 61
Warkentin, J. 111, 112
WBGU 49
Weld, C.R. 114
Westra, L. 30
whaling 25
wildlife 171, *173*, 174
women: Amerindians 183–4, *184*;
    Nigeria 125, 133, 134–6, 137;
    Southern Africa 144, 148–9, 151
World Bank 53
World Biosphere Reserves 49
World Commission on Environment and
    Development (WCED) 158
World Health Organisation 82;
    European Centre for the Environment
    and Health 87; European Charter on
    Environment and Health 87; Health
    for All 2000 86, 87; Healthy Cities
    project 82–4
World Resources Institute (WRI) 54–5
World Trade Organisation 53

Zamuteba 162
Zarewa 125, *127*, 128, 130, 131–2,
    133–7
Zimbabwe: economy 147; education
    145, 152; employment 153; land
    availability 146, *see also* Mzunga